The Social Side of Tourism: The Interface between Tourism, Society, and the Environment

SCHRIFTENREIHE DES IMT 9

Schriftenreihe des Instituts für
Management und Tourismus

Herausgegeben von Christian Eilzer,
Bernd Eisenstein und Wolfgang Georg Arlt

PL ACADEMIC RESEARCH

Michael Lück / Jan Velvin / Bernd Eisenstein (eds.)

The Social Side of Tourism: The Interface between Tourism, Society, and the Environment

Answers to Global Questions from the International Competence Network of Tourism Research and Education (ICNT)

PL ACADEMIC RESEARCH

Bibliographic Information published by the Deutsche Nationalbibliothek
The Deutsche Nationalbibliothek lists this publication in the Deutsche Nationalbibliografie; detailed bibliographic data is available in the internet at http://dnb.d-nb.de.

Library of Congress Cataloging-in-Publication Data
The social side of tourism: the interface between tourism, society, and the environment : answers to global questions from the International Competence Network Of Tourism Research and Education (ICNT) / Michael Lück, Jan Velvin, Bernd Eisenstein (eds.). — 1 [edition].
pages cm. — (Schriftenreihe des Instituts für Management und Tourismus (IMT), ISSN 2194-0002)
ISBN 978-3-631-66416-2
1. Tourism. 2. Tourism—Social aspects. 3. Visual communication—Research. I. Lück, Michael, 1966- editor. II. International Competence Network of Tourism Management.
G155.A1S583 2015
306.4'819--dc23

2015007528

ISSN 2194-0002
ISBN 978-3-631-66416-2 (Print)
E-ISBN 978-3-653-05558-0 (E-Book)
DOI 10.3726/978-3-653-05558-0

PL Academic Research is an Imprint of Peter Lang GmbH.

Peter Lang – Frankfurt am Main · Bern · Bruxelles · New York · Oxford · Warszawa · Wien

This publication has been peer reviewed.

www.peterlang.com

Foreword

In 2007, colleagues from a number of universities around the globe were invited by the West Coast University of Applied Sciences in Heide, Germany to attend a conference to celebrate the launch of a new collaborative network of likeminded people. In subsequent months and years, this initiative was enthusiastically supported, and eventually given the name International Competence Network for Tourism Research and Education (ICNT). Annual conferences and AGMs have taken place at partner universities ever since, such as Mexico City (Mexico), Potchefstroom (South Africa), Auckland (New Zealand), Drammen (Norway), Kamloops/Sun Peaks (Canada) and Porvoo (Finland). While the collaborative interest is great, specialisations and topic areas of research vary significantly from member to member. Despite this diversity, common grounds are always found, and the conferences result in subsequent publications, such as a volume on national park tourism (Eilzer et al., 2008) and global experiences in tourism (Eilzer et al., 2011). This book is the third in this series, and once again ties together a number of tourism experts, with a wide range of interest and expertise.

The overall objective of this third volume has been to increase the understanding of challenges in how tourism is understood and worked with in different countries around the world, with a focus on the interface between the social, the natural, and the economic environments. An edited book like this always builds on many individual contributors' ways to see tourism as a driving force for improving livelihoods of residents in both urban and rural regions. As editors we aimed to bring together research from different parts of the world that may boost the understanding of the tourism industry, and the composition and different study approaches of various tourism business activities. Such activities, described in the book, contain issues at both strategic and operative levels.

The underlying bottom line is how to increase competitive advantages, and how tourism could be a part of the local value creation. This includes social and cultural perspectives, both of host communities and of tourists, as demonstrated in the first section. Section Two contends that nature, and the preferences for use of nature, are at the forefront of the visitor experience, especially in protected areas. Lastly, Section Three offers insight into the conditions for the branding of tourism destinations, as well as tourism strategies for local community economic development.

The relationship of tourism and place appears to be particularly important in several of the research questions answered in this book. In order to ensure a demand for different types of tourism, it also requires an understanding of the desires and needs of various customers and providers, and their relationship to nature is also presupposed. The more time tourists spend at a place, the larger the demand for products and services from local providers may be expected. Subsequently, the development of experiences and services demanded by tourists are an important foundation for obtaining the objective of continuous settlement and employment, especially in rural areas.

With the very diverse backgrounds, and indeed research focus and expertise of the contributing authors, this book provides a kaleidoscope of research in the wide field of tourism studies. Such diversity could potentially endanger the coherence of such a volume; however, in this case it is seen as a genuine strength of the book. We applaud the authors' genuine concern for the host communities, tourists, and the environment. Their work will bring us one step closer to a sustainable tourism, supported by strong communities and outstanding tourist experiences.

We would like to take the opportunity to thank all ICNT members, for hosting and attending the annual meetings, and for their continuous support. In particular, we are indebted to the authors, who have generously contributed to this book: Kristian Bjørnstad, David Di Tota, Eva Holmberg, John Hull, Alexander Koch, Anne Köchling, Martinette Kruger, Pascal Languillon, Kaija Lindroth, Tomas Pernecky, Brooke Porter, Jarmo Ritalahti, Melville Saayman, Jonathon Spring, Xiaoshu Sun, and Peet van der Merwe. Your contributions made this book a valuable source for students, researchers, and practitioners alike.

Michael Lück, Jan Velvin and Bernd Eisenstein

References

Eilzer, C., Arlt, W.G. & Eisenstein, B. (eds.) (2011). *Global Experiences in Tourism. Proceedings of the International Competence Network of Tourism Management (ICNT).* Munich, Martin Meidenbauer Verlag.

Eilzer, C., Eisenstein, B. & Arlt, W.G. (eds.) (2008). *National Parks and Tourism. Answers from the International Competence Network of Tourism Management (ICNT).* Munich, Martin Meidenbauer Verlag.

Table of Contents

Social and Cultural Dimensions of Tourism

Tomas Pernecky

Tourism, Prejudice and Societal Conflict

1. Introduction

The 21st century continues to be marked with disagreement, violence, and lack of respect for human life. According to the United Nations Security Council, by the end of 2010, approximately 27.5 million people were displaced, with an additional 15.4 million becoming refugees due to conflict and violence (*Report of the Secretary-General on the protection of civilians in armed conflict* 2012). The World Development Report (*World development report 2011: Conflict, security, and development*, 2011) offers further insights on contemporary conflict, stating that over 1.5 billion people live in countries affected by violence, fragility or conflict; that only a few countries have a true 'post-conflict' status because violence is likely to recur; that the gap in poverty between countries impacted by violence and others is increasing and; that no low-income fragile or conflict-affected country has yet achieved a single United Nations Millennium Development Goal. With regard to the last point, the United Nations specified eight Millennium Goals in 2000, which include eradicating poverty and hunger, providing universal education, promoting gender equality, reducing child mortality, improving maternal health, combating diseases such as HIV, ensuring environmental sustainability, and developing a global partnership.

In contrast to these statistics is tourism – a phenomenon that can transform places into tourism destinations, give objects a touristic meaning, and turn people into hosts, maids, tour guides, waiters and tourists. This socially constructed machinery engages millions of people, and is capable of generating employment, revenues, and business activity. The question is whether it can make a difference in the communities that need it most. Tourism is understood to have both positive and negative aspects, and most undergraduate students should be able to articulate the ways in which tourism can benefit different stakeholders, while recognising also that there is a wealth of less desirable facets. For example, one only needs to browse an introductory text to learn about the negative impacts such as crime, noise levels, violence, pollution, drug and alcohol consumption, gambling, increase of cost of living, the destruction of local ecosystems, the loss of traditional values in host communities, changes in moral conduct for the locals, and much more (Rodgers, 2001). However, when presented with the more momentous questions, and situating tourism within the milieu issues noted in the World Development Report above, it can be a bewildering exercise for students and researchers alike to grasp the so-called 'hopeful' or 'peace-making' agency of tourism.

The premise of this paper is therefore to critically engage with the possibilities and limitations of tourism with regard to societal conflict, a topic that ought to be better integrated in tourism education, and to contribute to the development of a broader theoretical agenda that scrutinises tourism against important societal problems. The specific contribution of this paper lies in unfolding a conceptual framework which combines motivational theory with destination status in order to examine *if* and *how* different types of tourism may play a (significant) role in bringing about a positive change. The stance from which this paper is written does not pre-suppose that tourism is intrinsically good or bad, and instead sees the need for more situated and context-specific knowledge of tourism. In other words, tourism is offered as a volatile phenomenon and subject to many variables – thus necessary to be studied with regard to contexts (social, political, historical), people, cultures, bureaucratic processes, stakeholders' motives – all of which jointly shape the environment in which tourism exists, and its agency to be a force for positive change.

The answer to questions of what tourism *can do* in regard to societal conflict therefore demands a hermeneutic treatment (i.e. methodological approaches which take into account socio-cultural-political contexts, locality, customs, types of conflict, and historicity), and will differ depending on a multitude of factors. By developing a framework that seeks to position tourism, and mass-tourism in particular, in the wider context of a destination status, it is argued that only some forms of tourism can be assigned agency as to positive change in a post-conflict setting, which can be a lengthy and challenging process. It is also put forward that much of touristic contact is motivated by commercial interest and not by improving inter-group relations of people in a conflict/post-conflict setting. With regard to the organisation of the text, the manuscript first begins by acknowledging related projects and the work of relevant special interest groups. The paper then moves on to examining literature on the contact hypothesis and motivational theory necessary to introduce new conceptual models that examine tourism's role in recovery from societal conflict.

1.1 Recent Developments in Tourism Studies

It is apposite to say that tourism academics are in healthy disagreement as to the limits and possibilities of tourism, with voices on each side of the continuum. On one end of the spectrum are contributions representing the more critical voices, and on the other are the visions of hopeful intellectuals. There are also scholars in the middle – both hopeful and critical – who see potential for tourism to be a tool for positive change, but are aware of the fact that tourism is not the solution to problems that are rooted in deep cultural, political and social divides between peoples. In addition to these are works concerned mostly with the practicalities of tourism and the testing of hypotheses, but frequently isolated from critical theory and thus open to criticism (see, for example, the work on Mt Gumgang by Kim, Prideaux & Prideaux, 2007; Kim & Prideaux, 2003). Taken as a whole, the importance of all of these contributions should not be diminished, as they jointly advance knowledge on the ways in which tourism manifests in different contexts. The following sections offer a brief review of the key developments that are of particular relevance to discourse on tourism and societal problems.

1.2 Tourism and Conflict

In conceptual terms, conflict occupies a vast area that is intertwined with various research foci, and is prone to disparity with regard to its examination in the context of tourism. There are numerous researchers whose contributions to the study of tourism and conflict are important but, due to the limited space available, not all can be noted. The works noted here are highlighted as they are aligned with the overall aim of this paper, which is largely concerned with societal conflict. Societal conflict is not only restricted to acts of violence, as it can be coloured by all types of disputes: be they racial, religious, economic, or ideological. It can play out in different ways, and stem from class differences, gender, age, ethnicity and other social factors (Marger, 2011, p. 3). If we were to include globalisation in the picture, societal conflict can be seen as 'a function of culture, caused by a dysfunction of societal communication' (Flor, 2002, p. 3). By further drawing on literature on conflict management and resolution, societal conflict can be fathomed as follows:

> Societal conflict is a universal phenomenon, intrinsic to the process of social change. It is inevitable so long as material and social resources are unequally distributed within society, and inequity is reflected in cultural, social and political relationships between groups (Cliffe & White, 2002, p. 45).

With this basic outline in place, it is first necessary to acknowledge that tourism can produce, reproduce, and maintain states of anxiety and tragedy. From one point of view, tourism is the platform for attacks motivated by ideological aims (Freyer & Shroder, 2007) – such as when the opportunities provided by mass tourism (easy targets, large number of victims, the perceived hedonism of tourists and publicity) are exploited by terrorist organisations, as humanity witnessed in the Bali bombings in 2002 and 2005 (Putra & Hitchcock, 2008). From a different point of view, tourism can be a reminder of hostility and judgment – here the September 11 terrorist attacks on the World Trade Centre in New York have manifested *vis-à-vis* tourism through prejudice, discrimination, religious profiling, and xeno-racism. Stephenson and Ali (2010, p. 236) use the term *islamophobia* to denote the 'dread or hatred of Islam and therefore fear of dislike of all or most Muslims'. Tourism is thus not immune to the ways in which conflict manifests in this day and age; it is intertwined with conflict, and has ramifications for many aspects of tourism.

The consumption of conflict in tourism is perhaps best described by the term 'dark tourism'. The concept of dark tourism, popularised recently by Lennon and Foley (2000), was built on people's interest in disaster, death and atrocity. The way it transpires in the tourism phenomenon is through tourists visiting sites associated with human suffering, warfare, and genocide. Smith, MacLeod and Robertson (2010) note that dark tourism is not entirely new and has been in fact researched under different conceptual cappings, such as 'thanatourism' (Dann & Seaton, 2002; A. V. Seaton, 1996), 'Black Spots' tourism (Rojek, 1993), and in the context of heritage of atrocity (Tunbridge & Ashworth, 1996). Some researchers (Causevic & Lynch, 2011; Lynch & Causevic, 2008) have used the term 'phoenix tourism' to underscore the importance of social reconciliation and urban regeneration in the process of destination development. Although only little is known about the problems of dark tourism or thanatourism management (T. Seaton, 2009), more research is starting to emerge on issues such as place identity, destination management, motivation, marketing and interpretation of such sites (see, for example, the recentlu edited volume by White & Frew, 2013). The problems of interpretation are of particular significance and are revisited later in the paper.

An important point to clarify is that, when speaking of societal conflict, the discourse is not only limited to research on tourism and war (Butler & Suntikul, 2013), or tourism and terrorism (Higgins-Desbiolles, 2007; Putra & Hitchcock, 2008; V. Smith, 2005). It is entangled with matters of control and power (see the work of Church & Coles, 2007; Elliot, 2004; Hall, 1994; D. Hall, 2007; Macleod & Carrier, 2010), and entwined with politics, ethnic issues, indigenous rights, and hegemony (Burns & Novelli, 2007). It must also include the concerns existing in the less developed nations. These have been, for example, noted by Harrison (2001), and recently explored by Cole (2008), who immersed herself in the setting of a poor remote community in East Indonesia to unveil the conflicts of tourism development, and the challenges that arise between the different stakeholders. Societal conflict is also linked with responsible tourism (Leslie, 2012), conservation issues (Spenceley, 2012) and poverty (C. M. Hall, 2007; Scheyvens, 2011).

It is impossible to tackle all of these issues in one paper and therefore the conceptual analysis presented here is limited to a large-scale perspective on the role different forms of tourism play – mapped against destinations that have experienced conflict. Conflict can be underpinned by different principles and Tornblom and Kazemi (2012) list 13 different categories, noting that each demands a tailored solution. This paper does not delve into such depths of typological distinction, and focuses broadly on inter-societal disputes that have impacted local communities and saw the use of military/political power, continued state of occupation, physical destruction and violence. In other words, this paper is mostly concerned with observable nation-state and inter-group conflict such as the war in Bosnia and Herzegovina, the ongoing conflict between Israel and Palestine, the civil wars in Africa and the continued tensions between North and South Cyprus. Whilst other issues such as conservation are acknowledged to fall into the category of societal conflict, its examinations are not part of the offered analysis.

1.3 Tourism and Peace Research

The tourism-peace nexus has occupied tourism academics for nearly 25 years, and has undergone several stages: from euphoria in the late 1980s, to scepticism and disbelief, to revival and new levels of interest in 2000, to renewed belief in tourism as a contributor to peace (Moufakkir & Kelly, 2010; Salazar, 2006). In 1988, the field witnessed a new direction when a conference titled 'Tourism – A Vital Force for Peace' was held in Vancouver, Canada. Several scholars (Goeldner & Ritchie, 2003; Jafari, 1989) have commented on the importance of this event, which is now understood to have triggered a wave of peace-oriented research, giving impetus also to the establishment of the International Institute for Peace through Tourism (IIPT, 1999). This was the beginning of a hopeful vision that would lead to presenting tourism as a 'peace industry', embedded in the belief that 'tourism properly designed and developed, has the potential to help bridge the psychological and cultural distances that separate people of diverse races, colours, religions and stages of social and economic development' (D'Amore, 1988a, p. 154). Under the IIPT banner, every traveller was potentially 'an ambassador for peace' (IIPT, 1999).

Apart from the International Institute for Peace through Tourism (IIPT), still active today, other organisations, underpinned by the vision that tourism may be able to tackle significant societal problems, had been formed. The Israeli-Palestinian Tourism Forum (IPTF), founded in 2004, represented tourism and hotel industry professionals, such as travel agents and hotel executives. The goal of this forum was to strengthen dialogue, advance incoming tourism to the region, and solve common problems (Tourism4Peace, 2008). The Tourism4Peace Forum, founded in 2005, was also an initiative by the Israel Hotel Managers Association and the Peres Centre for Peace, with the aim to 'advance peace through tourism by strengthening dialogue and ties in the region' (Tourism-4Peace, 2008). The chief purpose of this endeavour was to encourage regional cooperation, ensure free movement of tourists between countries throughout the region, and organise joint marketing activities.

On the academic front, researchers have been optimistic about the possibilities of tourism also, and several scholars have upheld the more noble visions of tourism, including those that propose that tourism is characterised by 'understanding, tolerance and human dignity' Goeldner (1989, p. 167); that tourism is the largest 'peacetime movement of people in the history of mankind' (Jafari, Pizam, & Przeclawski, 1990, p. 469); that through travel we 'evolve a mutual trust and respect for one another and the dignity of all life on earth (D'Amore, 1988a, p. 154), find 'friends in every corner of the earth', and spread 'messages of hope for a peaceful world' (D'Amore, 1988b, p. 270); and that 'people engaged in any dimension of the vast tourism, hospitality, recreation and sports industries seek to provide a welcoming experience for travellers, much like the peace-making activities experienced within family: offering information, food, shelter, comfort and relaxation' (Haessly, 2010, p. 13). Researchers have also studied student perceptions and argued that tourism is a force for peace (Var & Brayley 1989; Var, Schluter, Ankomah, & Lee, 1989).

On the other side of the spectrum are those who posits that 'scholars have good reasons to be sceptical about the way the peace-through-tourism discourse is currently framed' (Salzar, 2006, p. 330). There are a number of academics who are not fully convinced by the tourism-for-peace proposition and its related claims (see for example Brown, 1989, p. 270; Cohen, 1972; Ferreira, 1999; Furnham, 1984; Gelbman & Maoz, 2012; Pizam & Mansfeld, 1996). In addition, a lot of critique is directed towards the Contact Hypothesis, discussed separately in greater detail in the following section. Then, somewhere in between, are the researchers who see the potential for tourism to be a positive force, but through a critical lens. In this regard, the Critical Toursim Studies (CTS) network, which has produced five conferences since 2005, gathers over two hundred academics who 'share a vision of producing and promoting social change in and through tourism practice, research and education' – as stated in the CTS V conference aims by the 2013 conference conveners (Minnaert et al., 2013).

Despite that, not all critical tourism studies academics carry out peace-related research; many delve into related societal problems including poverty, injustice, oppression, coruption, ecological crisis and so forth. Some, inevitably, write on conflict as noted in the previous section, hence the separation of scholars into clusters of peace and conflict is only artificial. Important to mention, under the critical voices, are also those that examine tourism's role in societies and the global community (Higgins-Desbiolles, 2006), and those who call for 'justice tourism' (Higgins-Desbiolles, 2008), 'hopeful tourism' (Pritchard, Morgan, & Ateljevic, 2011), and a 'moral turn' in tourism studies (Caton, 2012). Importantly, the examination of what is good and what outght to change about tourism, continues to attract scholarly activity. Since the publication of Tourism Progress and Peace (Moufakkir & Kelly, 2010), there is now a second edited volume by Blanchard and Higgins-Desbioles (2013), which offeres a collection of the latest critical essays on matters of tourism and peace.

2. Understanding the Limits and Possibilities of (Mass) Tourism

2.1 Tourism and the Contact Hypothesis

The study of prejudice has been greatly inspired in the early 1950s by Allport's (1958) landmark book The Nature of Prejudice. Allport developed a theory which sought to reduce prejudice and improve relationships between groups that are experiencing conflict. This came to be known as the 'contact hypothesis'. Allport's legacy is undeniable, for his ideas continue to stimulate academic debate across social sciences (Dovidio, Glick, & Rudman, 2008). The contact hypothesis is not a silver bullet and is subject to certain conditions. In order for the contact hypothesis to work (i.e. to change the negative attitudes between two groups experiencing conflict), Allport believed that there are certain conditions that have to be met. These are equal status between the groups; personal and sustained communication between individuals from the groups, cooperative interdependence (this is necessary for effective contact and involves cooperative activities), and the necessity for relevant authorities to endorse social norms in support of equality (Ellis & Maoz, 2008).

In theoretical terms, the contact hypothesis offers a promise to bridge the divides amongst people, bring one closer to another, and lead to a positive change in attitude. Furnham (1984, p. 51) states that studies on nationality 'have supported a culture-distance hypothesis, which suggests that the amount of difference between cultures (in terms of geography, religion, language) is directly proportional to the adjustments of that culture'. In the context of tourism, the contact hypothesis has been promoted as the 'major theoretical foundation for peace through tourism' (Kelly, 2012, p. 32). Tomljenovic (2010) confirms that the arguments of the tourism-peace nexus are generally underpinned by three main assumptions. The first two have origins in contact theory of social psychology and go as follows: first, tourism brings people together and thus provides opportunities for contact; second, this touristic contact is sufficient enough to result in mutual liking and greater understanding of the other. The third assumption is that this will lead to world peace.

Tomljenovic (2010) also points out that several of these assumptions have been contested. For example, tourism activities do not necessarily provide sufficient contact opportunities in order to become familiar with local cultures (Cohen, 1972). Furnham (1984) argues that tourists are often engaging in only certain type of social encounters that are designed for tourism, such as shopping, eating, drinking, and leisure and entertainment related activities. He further notes that the reaction of tourists to places differs from person to person: from being enchanted, delighted, adjusted in a new (touristic) setting as well as being bewildered, tense, and unhappy. Noteworthy is also the disparity between locals and tourists, which can lead to unfavourable responses towards tourism, particularly in societies with colonial past, and where tourism development can be met with aggression against tourists, feelings of resentment, feelings of envy, and the 'colonists' contemptuous and disparaging behaviour' (Dogan, 1989, p. 221).

In addition, large numbers of tourists have been noted to have a 'detrimental' effect on local cultures and customs (Cohen, Yeshayahu, & Almagor, 1992, p. 229), and it cannot be ignored that the tourism industry is driven by profit and can raise, not diminish, barriers between tourists and hosts (Bruner, 1991; Cohen, 1972; de Kadt, 1979; Nettekoven, 1979, cited in Tomljenovic, 2010). Tourism can lead to intergenerational conflicts, disruption of intimate and personal relations, and the transformation of relationships into a source of economic gain (Dogan, 1989). Indeed, a vast part of the travelling population is motivated by the desire to simply escape the daily reality; it has been acknowledged that tourism is motivated by hedonic reasons (Goossens, 2000; Mannell & Iso-Ahola, 1987; McCabe, 2000). Unequal economic status and superficial interactions between hosts and tourists can also pose a problem (Nyaupane et al., 2008). Not all touristic encounters are therefore meaningful, and experiencing different cultures can go either way.

Perhaps for all of the above reasons, it is not surprising that empirical studies on contact hypothesis in the field of tourism studies have yielded mixed results. A few selected works ought to be noted. In the examination of the effect of encounters between Jewish-Israeli tourists and Arab Egyptian hosts, Maoz (2010) found that, while positive attitude change with regard to stereotyping and prejudice was noticeable, a significant percentage of the sample also experienced negative attitudes towards the other group, which further intensified as a result of the contact. The findings of a different study on 97 Greek tourists travelling to Turkey for the first time, showed mostly negative perception of the host community (Anastasopoulos, 1992). Other studies, such as by Tomljenovic (2010), confirm increased cultural understanding through tourism only tentatively, and although there are some undeniable benefits to being exposed to different cultures through tourism, noted elsewhere in the literature, it has become clear that not all touristic contact reduces prejudice.

Many of these issues were in fact acknowledged by the 'father' of the contact hypothesis Allport (1958, p. 277), who proclaimed that 'it would seem fair to conclude that contact, as a situational variable, cannot always overcome the personal variable in prejudice'. He also warned about the nature of superficial contact, which may strengthen one's already existing mental associations, and make people (tourists) sensitised to seeing signs that confirm stereotypes. On this front, Pernecky (2010) has posited that tourists travel with their own 'baggage' and are shaped by different socio-cultural contexts. Perhaps the most robust critique of contact theory is offered by Forbes (1997) who demonstrates that increased contact between cultures can produce even more intense conflict but, despite that, he does not reject the notion that individuals who get to know each other better, in general, increase the chances of liking one another. The important question, then, concerns the kind of contact necessary to bring people in conflict closer.

2.2 Casual Contacts, Acquaintances and Goodwill Contact

The distinction between different types of contact, which can be found in Allport's (1958) original work, are a good starting point for developing a more comprehensive model to understand the ways in which (mass) tourism and various forms of touristic contact manifest in different settings. Allport explains that 'casual contacts' do not necessarily dispel prejudice, and can in fact increase it. Therefore meeting or seeing someone that is perceived as different or an outsider (e.g. minority groups, or people of different ethnic backgrounds) does not warrant effective communication and a change in attitude. In fact, stereotyping and pre-conditioning may lead to evasive, cold and distrustful behaviour. Therefore, depending on the specificities of the touristic encounter, a fleeting visit to a place and a superficial contact with locals may eventuate in a positive experience, but it can also reinforce stereotypes: such as "he is a Jew, he will skin me", or on the part of the host: "He is Irish, they hate Jews, he is likely to insult me" (see Allport, 1958, p. 252 for more information).

Where Allport saw more potential were contact situations that could lead to developing friendship or 'acquaintance' between groups, the reasons being newly gained knowledge and experience about the other, which can lead to a truer set of beliefs, and yield more friendly and tolerant attitudes. In other words – a deeper, personal process of questioning one's pre-disposed beliefs about the 'other' through a meaningful contact. Allport (1958, p. 255) held that 'contacts that bring knowledge and acquaintance are likely to engender sounder beliefs concerning minority groups and for this reason contribute to the reduction of prejudice'. It is worthy to note that, interestingly, case studies on tourism contact neither fully reject this notion nor do they confirm it. For example, a controlled study of an MBA Business Study Mission to the Middle East by Litvin (2003) did find a shift in pre-trip and post-trip attitudes, but in both directions. After visiting one destination (Egypt), the students' attitude changed in a negative direction, while positive changes were recorded after visiting a different destination (Israel).

'Goodwill Contacts' are described by Allport as initiatives to combat prejudice. Although Allport's attention was mainly directed toward inter-rational contact and minority groups – he was not concerned about tourism per se – goodwill contacts pertain to representatives of the community, and include minority groups. In the context of tourism, we may think of goodwill contacts as a form of tourism that is purposefully designed to reduce prejudice. The tourism initiatives noted earlier in this paper, such as the Israeli-Palestinian Tourism Forum (IPTF) and the Tourism4Peace Forum, are good illustrations of such efforts. There are also other forms of tourism such as pro-poor tourism, and a range of social entrepreneurs who strive to make a difference in the lives of local communities (see for example African Pro-Poor Tourism Development Centre, 2013). The obvious necessity for tourism researchers studying conflict and contact theory, is to make a distinction between casual contacts, goodwill contact and acquaintance, as all involve different types of actors. In order to eradicate potential misinterpretation of the agency of tourism, which ought not to be approached as a homogenous unit, Figure 1 outlines several important considerations.

Figure 1: Factors Impacting Touristic Encounters: In-between Contact and Conflict

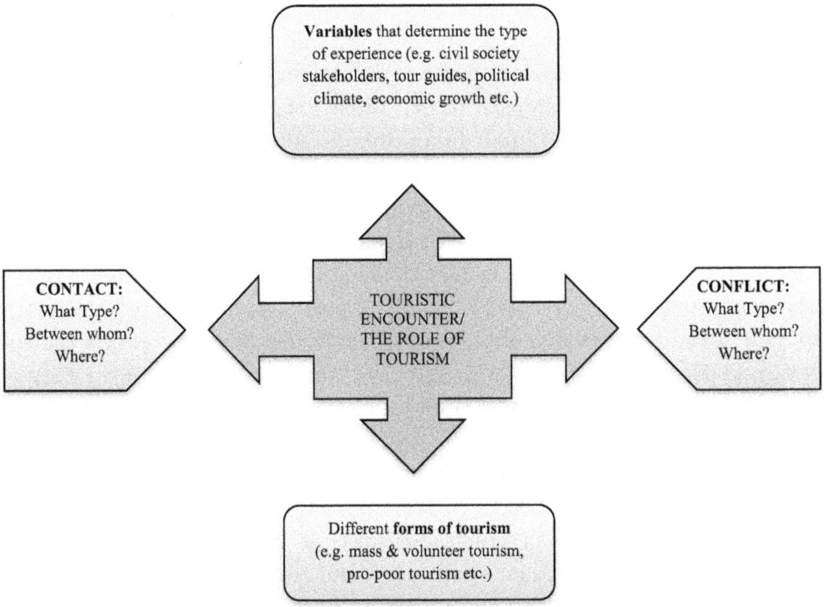

Figure 1 suggests that it is important to distinguish between various types of conflicts and contact, and that there are other factors that influence the nature of the touristic encounter. With regard to the notion of variables, academic enquiries into the possibilities of tourism in places with a post-conflict status represent a very promising area, and research in this area is starting to expose the complexities that come with tourism development in destinations such as Bosnia and Herzegovina. Causevic and Lynch (2011), for instance, explain that before the war, Bosniaks, Croats and Serbs lived together in a multicultural society, without the notion of 'ethnically clean' territories. The task at hand for the tour guides today – in a post-conflict country – is different and more challenging, for they need to engage in a complex process of interpretation. The expectations for tour guides to be mediators or 'brokers between the tourists' culture and the imagined local culture' (Salazar, 2006, p. 327), however, are problematic and pose some pitfalls. One of these is noted in a study by Gelbman and Maoz (2012), who focus on the messages the Israeli guides deliver to Israeli tourists at a place called Island of Peace (a border site between Israel and Jordan). The authors reveal that the information offered to tourists does not endorse peace; on the contrary, tourists are exposed to narratives of conflict and war.

The role of guides as interpreters of local cultures is therefore a significant contender for consideration in relation to the study of prejudice. There are, nevertheless, other drawbacks that impact the nature of the experience for the tourist. By taking a part in an organised tour, and with interpreters on hand, tourists are claimed to be sheltered from having to interact with locals (Furnham, 1984), and the overall experience, which can indeed be educational, may provide only a superficial contact with the local people. Moreover, taking tourists to heritage sites has been argued to inhibit mutual understanding, as these places are political resources that 'aim to legitimise a specific social reality which divides people into 'we' and they' (Poria & Ashworth, 2009, p. 522). Noy's (2011) critical ethnographic work in the City of David, in East Jerusalem, exposes how tourism serves political objectives, and he presents a myriad of other issues, such as power, hegemony, site naming, authority and authenticity. Additional challenges faced by a post-war country are identified by Causevic and Lynch (2012), who underscore the intricacies of socio-political settings in which tourism is deeply rooted. Similarly, Scott (2012) emphasises the need for civil society stakeholders and higher level formal government mechanisms to support tourism initiatives. What therefore appears to be a fundamental component in understanding tourism's role in societal conflict, are the many variables, which must be interwoven into the analysis and tourism development.

The benefits associated with tourism to assist with economic growth, offer and promote business and social opportunities to encourage interaction among residents, and to 'accelerate the healing process' in post-conflict societies, are not always easily implemented (Novelli, Morgan, & Nibigira, 2012). Not only is there a vast difference between destinations, there are pitfalls specific to poor and low-income countries. Noveli, Morgan and Nibigira (2012) stress that places such as sub-Saharan Africa have to deal with many challenges. These include definitional confusion, unrealistic expectations and inadequate guidance over practical implementations in tourism development. Obstacles of this nature hinder the ability of tourism to deliver sustainable and equitable growth. There are also differences with regard to the type and scale of conflict and the destination under consideration. Tourism 'as a pathway to socio-economic recovery' is not a straightforward process (Novelli et al., 2012, p. 1449) – it requires cooperation among former victims and perpetrators which, of course, can be a challenging vision to achieve.

So far the discourse has focused on the contact hypothesis and the role tourism may play in shifting people's attitudes, but there is a conceptual gap when it comes to understanding the betterment of relations *vis-à-vis* tourism from a larger perspective. Before taking some of these issues further, it is necessary to examine the role of mass tourism in delivering hopeful prospects for societies in need. In this regard, Figure 1 makes a case for acknowledging the disparity between various tourism phenomena by noting that there are different forms of tourism, which determine the type of touristic encounters. Large cruise-ships, low-cost airlines and major tourism operators – all engaged in moving millions of people around the world – are poles apart from the small percentage of pro-poor tourists and activist groups. In other words, it is important to make a distinction between the different forms of tourism to appreciate which forms of tourism make what impact. To attain a greater sense of the possibilities mass tourism beholds as an agent of positive change, the application of established motivational frameworks will assist in piecing together a more pragmatic picture.

2.3 Re-visiting Motivation Frameworks: The Possibilities and Limits of Tourism

As already established, the contact hypothesis is underpinned by claims that tourism brings people together, increases contact and, therefore, also understanding and mutual liking amongst peoples. In order to determine the plausibility of this prospect, it pays to further examine the extent to which peoples' attitude changes (positively) through the act of travelling, as well as their willingness to, and reasons for, visiting certain destinations. Making sense of tourist behaviour and motivation has preoccupied several scholars, and the field has been enriched by a number of tourist typologies and motivational theories (for example, Cohen, 1972; Hamilton-Smith, 1987; Pearce, 1988; Plog, 1974, 2001; Redfoot, 1984). Such typologies are important and valuable theoretical tools, for they can bring more clarity into the relationship between different forms of, and societies/destinations in, conflict. Namely, they can tell us about the type of contact mass tourism facilitates (e.g. with regard to meaningful encounters and intercultural understanding), and also cast more light on where the majority of touristic contact takes place (e.g. with regard to tourism destinations).

When it comes to the psychographic profile of tourists, the seminal work of Stanley Plog (1974, 2001) has shown that vast differences can be observed. On the one hand, there are what Plog calls the *Psychocentrics/Dependables* who are cautious, conservative, and restrictive in spending; prefer popular and well-known products/destinations; are more lethargic and passive; tend to retreat in difficult situations; look to authority to make judgements; like structure and routine; and prefer the company of family and friends. On the other side of the spectrum, there are the Allocentrics/Venturers, who are more intellectually curious, tend to make quick decisions, spend income more readily, like new products/destinations, are self-confident, make their own judgement, are active and assertive, and prefer activities and challenges (Plog, 2001). The implications for travel in general are that Dependables travel less, spend less, go to highly developed 'touristy' spots, chose familiar recreational activities, prefer 'sun and fun' destinations, and can be classified as infrequent international tourists who tend to visit best-known places (Plog, 2001).

What Plog's work does well is demonstrate that it is the aspects of personality that determine where and why people travel. In this regard, contemporary research (Chen, Mak, & McKercher, 2011, p. 126) confirms that travel personality continues to be important in choosing a destination. Figure 2, adopted from Chen, Mak and McKercher (2011), combines some of the established works on tourist behaviour (Cohen, 1972; Pearce, 1988; Plog, 1974, 2001) which can be applied to examine the conceptual footing of the tourism-conflict nexus. What can be argued rather persuasively, is that if a place is not well-known, safe and, more importantly, is not perceived as safe, which is a lengthy process in terms of gaining the confidence of tourists, the vast majority of travelling public will avoid it. This is important to understand, as it tells us about the kind of contact and cultural exchange mass tourism provides. By bringing in the issue of safety and destination status, we can see where and among whom tourist encounters are likely to materialise.

Figure 2: Integrated Motivation Framework and Tourist Behaviour Pattern

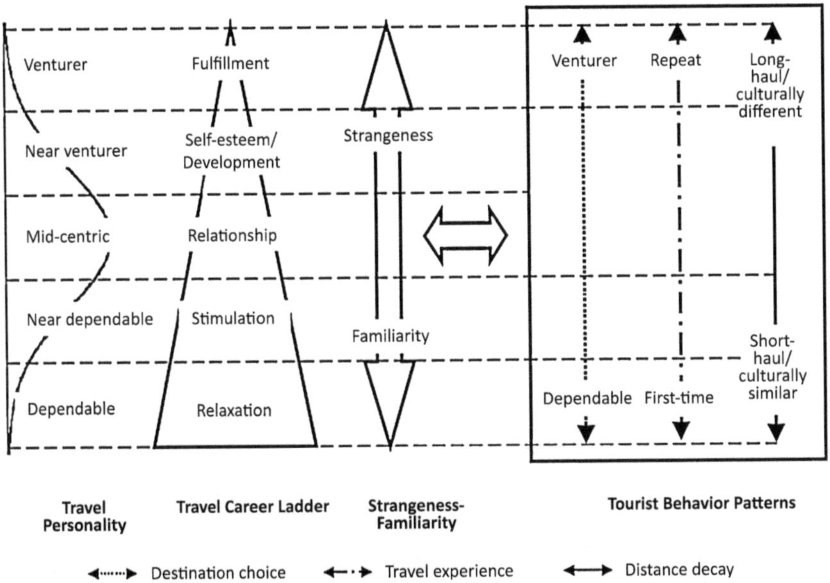

Source: Adopted from Chen, Mak and McKercher (2011, p. 127).

By making a distinction between destinations that are in-conflict, and places that can be assigned a post-conflict or no-conflict status (in terms of long term stability), it is possible to derive a more realistic view of the potential for tourism. Figure 3, following, shifts the focus to the destination status, to emphasise that no tourism is present in destinations that are in conflict, and that the majority of touristic activity occurs in regions that are safe, and seen as stable with regard to the post-conflict status. What it also highlights is that most tourists, who could hypothetically take part in making a difference to societies and places in need, in fact do not. Figure 3 suggests that the recovery process of a post-conflict destination can take a long period of time before it gains the confidence of tourists and is perceived as safe. The length of time this may take is subject to many variables, and the recovery process is likely to vary depending on the destination and the scale and type of conflict. The purpose of the figure, therefore, is not to accurately predict when mass tourism arrives, rather it is to illustrate the relationship of the different forms of tourism in regard to conflict.

Figure 3: Tourist Behaviour Contrasted with Destination Status

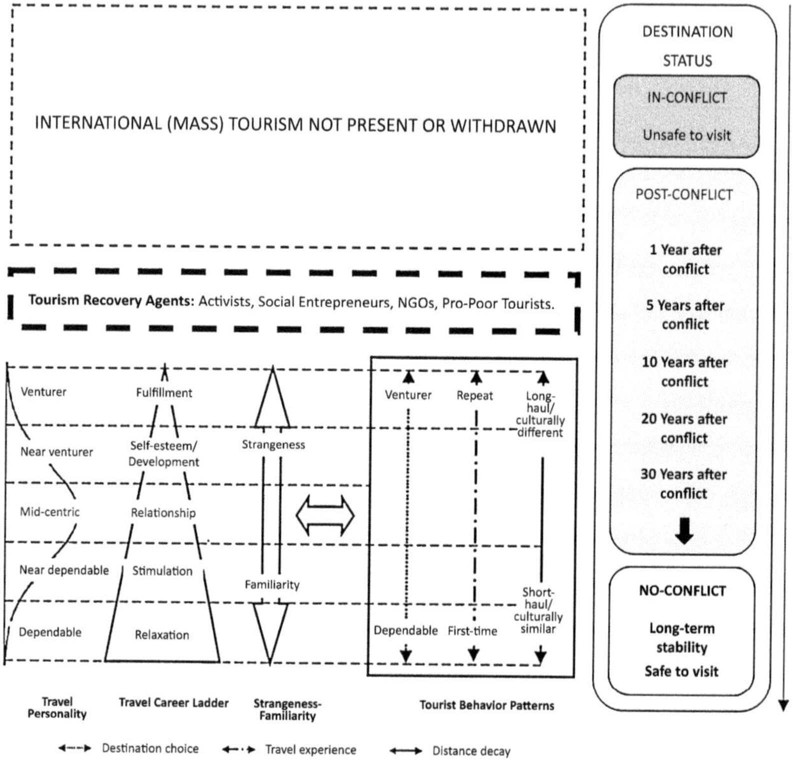

This figure depicts the role of tourism with regard to destination status (i.e. in-conflict, post-conflict and no-conflict). It suggests that only a small percentage of tourists identified by Plog (2001) as 'Venturers', and those that go beyond this label–committed to making a difference and noted here as tourism recovery agents – may be in a position to positively contribute to countries that are recovering from conflict. This figure also shows that it is a long process before post-conflict destinations can benefit from a large number of tourists and some of the positive aspects associated with tourism. Caveat: The number of years (e.g. 1-5-10-20-30) in the post-conflict box is not fixed and will vary depending on the type of destination and other variables, such as economic slowdown, political stability in the region, and the degree and complexity of conflict.

The fact that tourism can neither cease nor prevent conflict, because it occurs in safe and conflict-free locales, is in accordance with the opinion of those researchers who have argued that tourism is a beneficiary of peace (Burkart, 1988; Litvin, 1998), and that the worldwide phenomenon of tourism is dependent on peace and the ability to provide a secure environment for the tourist (Ferreira, 1999; Pizam & Mansfeld, 1996). This realisation needs to be underscored, because the agency of tourism is minimal when it comes to existing conflict; currently there are no studies that can show otherwise. On this note, one of the limitations of Figure 3 is that it does not make a distinction between different types of conflict. Without a doubt, the Bosnian War is different to massacres, violence and civil wars in Africa, which is different again to ongoing tensions between Northern and Southern Cyprus. The type of conflict and the contexts within which conflict unfolds is important to understand, as noted in Figure 1. Notwithstanding, the factor which determines tourism development, and the choices made by mass tourists, is destination safety and security – a precondition intertwined with political regimes (Lisle, 2007; Scott, 2012).

Places that are perceived through the lens of conflict on the part of the travelling public, media and tourism organisations, are not the type of destinations mass tourism and the Centric-to-Dependable typologies of travellers typically visit. A different point altogether, is the potential for tourism to reduce or negate prejudice. In this regard, even places that continue to operate and attract tourists, have not demonstrated a significant shift with regard to reconciliation of peoples. For example, the growth of tourism in Southern Cyprus has not (yet) yielded better relations between the Northern and Southern Cypriots. As a matter of fact, the flow of nationals and tourists is heavily controlled – limiting contact opportunities between peoples of the nations in dispute (see Entry Regulations for Cyprus, 2013). In addition, recent research shows that reconciliation in Cyprus through tourism has thus far led to very little progress (Scott, 2012), and that existing (small-scale) collaborative tourism partnerships have produced minimal success due to issues of reciprocity, and the fact that the commercial imperatives of large tour operators are not driven by interest in peace.

This brings us to the question on between whom touristic contact ought to occur (see Figure 1) for tourism to be able to reduce prejudice. Thus far it has been established that mass tourism is mainly present in conflict-free destinations and that, depending on the nature of the tourist's experience and a multitude of other variables, touristic encounters may both re-affirm negative, and alter positively, pre-existing perceptions of the 'other'. Returning to Plog's (2001) typology, the field of tourism studies already recognises that there are tourists who are more open to exploring new places and willing to learn about others cultures – these are Plog's Ventures. What has not yet been articulated with sufficient clarity in the literature, and what Figure 3 strives to emphasise, is that the remaining typologies such as Centric-Venturers, Near-Dependables and Dependables, largely visit well-known and safe places such as Paris, Las Vegas and Honolulu. Hence, a large portion of tourists avoid places with a post-conflict status, and so far there is no theory to support the claim that mass tourism at large reduces prejudice.

Conceptually speaking, Figure 3 does make a provision for Plog's 'Venturers' – aligned by Chen, Mak and McKercher (2011) with people's desire for fulfilment, and willingness to undergo less familiar experiences. In Plog's (2001) framework, these are tourists who prefer exploring independently new destinations, embrace challenges, and make their own decisions. Nevertheless, as shown in Figure 3, this segment still seeks safe places, as suggested by Plog's list of destinations (e.g. Vietnam, Fiji, Tibet, Nepal etc.), and places that are relatively underdeveloped and unspoiled by mass tourism (e.g. Antarctica and Amazon). Apart from Venturers, it is vital to acknowledge and accommodate the forms of tourism that are intentionally striving to improve the livelihoods of people in post-conflict destinations. Figure 3 uses an additional category called here *tourism recovery agents* to mark this difference. The category of *tourism recovery agents* goes beyond the label of Venturers (e.g. backpackers), and points to the specific types of tourism such as Pro-poor Tourism (see, for example, the work of Higgins-Desbiolles & Blanchard, 2010), the emerging strain of Ethical Tourism (see, for example, Tourism Concern, 2012), and other groups aiming to advance social innovation and encourage inter-group discussion and reconciliation. These can be conceived of in terms of Allport's (1958) notion of Goodwill Contacts.

By way of summing up the overall observations from the above depictions, several points can be made: One, peaceful relations between destinations and security are pre-requisite for mass travel to occur; Two, for a destination to benefit from mass-tourism and its economic impacts, tourists need to perceive it as safe and well-developed (in terms of the tourism product, infrastructure and so forth); Three, not all tourists are motivated by meaningful encounters and the desire to learn about other cultures; Four, claims suggestive of a tourism ability to reduce conflict and prejudice are yet to be demonstrated as the vast majority of tourists avoid places of conflict; Five, tourists motivated by relaxation and familiarity are not likely to visit places where they may be confronted by poverty and cultures and customs they are not familiar with, and; Six, tourists motivated to make a difference in the lives of others, noted in this paper broadly as *tourism recovery agents*, represent a small percentage of the travelling population, and their contribution as far as the reduction of prejudice and societal conflict is concerned, is a matter for further research.

In addition, whilst there are forms of tourism motivated to help communities in need, it can be argued that the people who visit post-conflict destination are already 'open minded', keen to learn and discover. The questions that arise, and where more research is needed, relate to understanding how a contact between *tourism recovery agents*, Venturers and host communities in post-conflict destinations is to reduce prejudice between the groups in dispute. There is no doubt that pro-poor tourism and volunteer tourism, for example, is important and can assist pockets of people impacted by conflict, but there is no evidence that this form of tourism reduces tensions between the countries and their peoples. In other words, these are two separate tenets that ought to be acknowledged in order to further progress research in this domain. In this regard, the questions offered in Figure 1 – *Contact/Conflict between whom?*, *Where?*, and *What type of contact/conflict?* – although simple, are fundamental for advancing knowledge on the possibilities of tourism to lessen societal conflict.

3. Conclusions

This paper set out to conceptually examine whether or not different types of tourism can significantly feature in bringing about a positive change. The aim was to understand the role of tourism in reducing societal conflict by taking a larger perspective, and employing well established theoretical models in the field of tourism studies. The paper has drawn on motivational theory and the work of Stanley Plog (2001) in order to cast more conceptual light on the relationship between mass tourism and destinations that were categorised according to conflict status. The offered findings purport that there is no empirical evidence to suggest that tourism can stop, prevent, or eliminate conflict from occurring. The majority of touristic activity transpires in regions that are safe, and seen by tourists as stable. In fact, it can be argued that (mass) tourism actively avoids destinations and communities living in conflict and that it can be a long process for tourism to flourish in post-conflict societies.

Conceptualisations of tourism that draw on the contact hypothesis are yet to be clearly formulated, in particular assertions entangled with tourism and peace propositions. While there is potential for tourism to be part of processes of reconciliation, normalisation and economic recovery, however, only when speaking of the hosting community, the current views that depict tourism as a positive force are prone to fallacy when presenting this type of touristic encounters under the banner of contact theory. A contact between tourists and hosts who may already have amicable relations (e.g. British tourists visiting Cyprus or Australians visiting Israel) is by no means correlated with improving relations among populations in conflict (e.g. between North and South Cypriots or Israelis and Palestinians). It would appear that such contact pertains only to some type of maintenance function of tourism, and is only of benefit to the hosting community.

For a contact hypothesis to 'work', even the founder of the contact hypothesis himself (Allport, 1958) acknowledged that superficial and casual contact is not likely to dispel prejudice and cure the problems of inter-group relations. Drawing on Allport's work, this paper argued that for contact to be meaningful and to result in positively changing one's attitude towards the previously prejudiced 'other', it needs to be at a level of acquaintance and friendship, and that it requires collaboration, equal status and support of relevant authorities. This is indeed one of the stumbling blocks in the study of tourism, as much of the touristic contact is known to be superficial and not driven by a desire to improve societal relations. Tourism can be, and is, used intentionally by different stakeholders. The contemporary realities of the vast majority of tourism development – the driving force behind mass tourism – is fuelled by commercial interests.

A mere contact neither warrants nor indubitably promotes better relations, and therefore does not always contribute to more positive understandings among people. In this regard, this paper has also highlighted that neither does tourism always provide the best setting for cross-cultural understanding. The complexity of the tourism phenomenon, together with the range of actors and variables that determine the tourist experience, requires tailored approaches to post-conflict tourism development. Furthermore, there are significant obstacles on the path of bettering justice-conflicts through tourism, such as undemocratic governments, and unless sufficient support is granted by relevant stakeholders, tourism initiatives, despite having good intentions, can be hindered (as noted by Scott, 2012 and others). The questions that should drive future research in this domain of inquiry must address the specificities of such encounters and take into account the construction of the tourist experience. If tourism is to reduce prejudice, it demands a far more comprehensive analysis of the variables that determine possible change in people's attitudes.

In order to end on a constructive, but also critical and hopeful, note, a suggestion of great relevance to the issues discussed in this paper has to be offered for instructors teaching on the subject of tourism and peace. The documentary Uspomene 677 (2011) follows the stories of teenagers of different ethnic backgrounds, who live in a post-conflict society in the aftermath of the Bosnian war. The documentary exposes the enormity of the task of reconciliation in a divided society, and should serve as a valuable resource for classroom discussions, as well as an inspiration for possible student-oriented projects. Finally, there are no qualms about the ability of some tourists in some situations changing their attitude about a culture and peoples in a more positive direction. There are instances where tourism can result in reducing stereotypes and prejudice, however, not all touristic contact is positive, and not all contact leads improving relations. This is not a given of tourism.

References

African Pro-Poor Tourism Development Centre (2013). Retrieved 17 May, 2013, from http://www.propoortourism-kenya.org/what_is_propoortourism.html.

Allport, G. W. (1958). *The Nature Of Prejudice* (Abridged ed.). Garden City, NY, USA: Doubleday & Company Inc.

Anastasopoulos, P. G. (1992). Tourism and attitude change: Greek tourists visiting Turkey. *Annals of Tourism Research, 19*(4), 629–642. doi:http://dx.doi.org/10.1016/0160-7383(92)90058-W

Blanchard, L.-a., & Higgins-Desbiolles, F. (2013). *Peace through tourism: Promoting human security through international citizenship.* London, UK: Routledge.

Brown, F. (1989). Is tourism really a peacemaker? *Tourism Management, 10*(4), 270–271. doi:10.1016/0261-5177(89)90001-0

Bruner, E. M. (1991). Transformation of self in tourism *Annals of Tourism Research, 18*(2), 238–250. doi:10.1016/0160-7383(91)90007-X

Burns, P. M., & Novelli, M. (2007). *Tourism And Politics: Global Frameworks And Local Realities.* Amsterdam, The Netherlands: Elsevier Science & Technology Books.

Burkart, A. J. (1988). Tourism — a vital force for peace. *Tourism Management, 9*(3), 254. doi:10.1016/0261-5177(88)90045-3

Butler, R., & Suntikul, W. (2013). *Tourism and War.* Abingdon, Oxon, UK: Taylor & Francis.

Caton, K. (2012). Taking the moral turn in tourism studies. *Annals of Tourism Research, 39*(4), 1906–1928. doi:http://dx.doi.org/10.1016/j.annals.2012.05.021

Causevic, S., & Lynch, P. (2011). Phoenix Tourism: Post-Conflict Tourism Role. *Annals of Tourism Research, 38*(3), 780–800. doi:10.1016/j.annals.2010.12.004

Causevic, S., & Lynch, P. (2012). Political (in)stability and its influence on tourism development. *Tourism Management*(0). doi:10.1016/j.tourman.2012.04.006

Chen, Y., Mak, B., & McKercher, B. (2011). What Drives People to Travel: Integrating the Tourist Motivation Paradigms. *Journal of China Tourism Research, 7*(2), 120–136. doi:10.1080/19388160.2011.576927

Church, A., & Coles, T. (2007). *Tourism, Power and Space.* London, UK: Routledge.

Cliffe, L., & White, P. (2002). Conflict Management and Resolution in the Horn of Africa. In C. Mwaûra & S. Schmeidl (Eds.), *Early warning and conflict management in the Horn of Africa* (pp. 43–68). Lawrenceville, NJ, USA: Red Sea Press, Incorporated.

Cohen, E. (1972). Toward a sociology of international tourism. *Social Research, 39*(1), 164–182.

Cohen, E., Yeshayahu, N., & Almagor, U. (1992). Stranger-Local Interaction in Photography. *Annals of Tourism Research, 19*(2), 213–233.

Cole, S. (2008). *Tourism, Culture and Development: Hopes, Dreams and Realities in East Indonesia*. Clevedon, UK: Channel View Publications.

Dann, G., & Seaton, A. V. (2002). *Slavery, contested heritage, and thanatourism*. London, UK: Taylor & Francis.

Dovidio, J. F., Glick, P., & Rudman, L. (2008). *On the Nature of Prejudice*. Malden, MA, USA: Wiley.

D'Amore, L. J. (1988a). Tourism – a vital force for peace. *Tourism Management, 9*(2), 151–154. doi:10.1016/0261-5177(88)90025-8

D'Amore, L. J. (1988b). Tourism – A Vital Force for Peace. *Annals of Tourism Research, 15*(2), 269–270. doi:10.1016/0160-7383(88)90087-4

de Kadt, E. (1979). The encounter: changing values and attitudes. In E. de Kadt (Ed.), *Tourism: Passport to development?* (pp. 50–67). Oxford, UK: Oxford University Press.

Dogan, H. Z. (1989). Forms of adjustment: Sociocultural impacts of tourism. *Annals of Tourism Research, 16*(2), 216–236. doi:10.1016/0160-7383(89)90069-8

Elliot, J. (2004). *Politics of Tourism: A Comparative Perspective*. London: Routledge.

Ellis, D. G., & Maoz, I. (2008). Dialogue, Argument, and Cultural Communication Codes between Israeli-Jews and Palestinians. In L. A. Samovar, R. E. Porter, & E. R. McDaniel (Eds.), *Intercultural Communication: A Reader* (pp. 244–251). Boston, MA, USA: Wadsworth Cengage Learning.

Entry Regulations for Cyprus. (2013). Retrieved 20 May, 2013, from http://www.cyprusembassy.net/home/index.php?module=page&cid=94.

Ferreira, S. L. A. (1999). Crime: A threat to tourism in South Africa. *Tourism Geographies, 1*(3), 313–324. doi:10.1080/14616689908721324

Flor, A. G. (2002). *Communication and Culture, Conflict and Cohesion*. Los Baños, Laguna, Philippines: UPLB College of Development Communication and The Foundation for Development and Communication.

Forbes, H. D. (1997). *Ethnic Conflict: Commerce, Culture and the Contact Hypothesis*. New Haven, USA: YALE University Press.

Freyer, W., & Schroder, A. (2007). Tourism and terrorism: An analytical framewrork with special focus on the media. In E. Laws, B. Prideaux, & K. Chon (Eds.), *Crisis management in tourism* (pp. 129–141). Wallingford, UK: CABI

Furnham, A. (1984). Tourism and culture shock. *Annals of Tourism Research, 11*(1), 41–57. doi:10.1016/0160-7383(84)90095-1

Gelbman, A., & Maoz, D. (2012). Island of Peace or Island of War: Tourist Guiding. *Annals of Tourism Research, 39*(1), 108–133. doi:10.1016/j.annals.2011.05.008

Goeldner, C. R. (1989). Tourism — a vital force for peace. *Tourism Management, 10*(2), 166–168. doi:10.1016/0261-5177(89)90058-7

Goeldner, C. R., & Ritchie, J. R. B. (2003). *Tourism - Principles, Practices, Philosophies* (9th ed.). Hoboken: John Willey & Sons.

Goossens, C. (2000). Tourism information and pleasure motivation. *Annals of Tourism Research, 27*(2), 301–321. doi:10.1016/s0160-7383(99)00067-5

Haessly, J. (2010). Touirsm and a Culture of Peace. In O. Moufakkir & I. Kelly (Eds.), *Tourism, progress and peace* (pp. 1–16). Wallingford, UK: CABI.

Hall, C. M. (1994). *Tourism and Politics: Policy, Power and Place*. Chichester: Wiley.

Hall, C. M. (2007). *Pro-poor Tourism: Who Benefits?: Perspectives on Tourism and Poverty Reduction*. Clevedon, UK: Channel View Publications.

Hall, D. (2007). *Tourism and Transition: Governance, Transformations and Development*. Oxford, UK: CABI Publishing.

Hamilton-Smith, E. (1987). Four kinds of tourism? *Annals of Tourism Research, 14*(3), 332–344. doi:10.1016/0160-7383(87)90106-X

Harrison, D. (2001). *Tourism and the Less Developed World: Issues and Case Studies*. Wallingford, UK: CABI Publishing.

Higgins-Desbiolles, F., & Blanchard, L. (2010). Challenging peace through tourism: Placing tourism in the context of human rights, justice and peace. In O. Moufakkir & I. Kelly (Eds.), *Tourism, progress and peace* (pp. 35–47). Wallingford, UK: CABI.

Higgins-Desbiolles, F. (2006). More than an "industry": The forgotten power of tourism as a social force. *Tourism Management, 27*(6), 1192–1208. doi:http://dx.doi.org/10.1016/j.tourman.2005.05.020

Higgins-Desbiolles, F. (2007). Hostile meeting grounds: Encounters between the wretched of the earth and the tourist through tourism and terrorism in the 21st century. In P. M. Burns & M. Novelli (Eds.), *Tourism And Politics: Global Frameworks And Local Realities* (pp. 309–332). Amsterdam, The Netherlands: Elsevier Science & Technology Books.

Higgins-Desbiolles, F. (2008). Justice Tourism and Alternative Globalisation. *Journal of Sustainable Tourism, 16*(3), 345–364. doi:10.1080/09669580802154132

IIPT. (1999). International Institute for Peace though Tourism Mission Statement. Retrieved August 21, 2012, from http://www.iipt.org/AboutUs.html

Jafari, J. (1989). Tourism and peace. *Annals of Tourism Research, 16*(3), 301–454. doi:10.1016/0160-7383(89)90059-5

Jafari, J., Pizam, A., & Przeclawski, K. (1990). A sociocultural study of tourism as a factor of change. *Annals of Tourism Research, 17*(3), 469–472. doi:10.1016/0160-7383(90)90011-F

Kelly, I. (2012). Peace through tourism: An implementation Guide. *The Journal of Tourism and Peace Research, 2*(2), 32–49.

Kim, S. S., & Prideaux, B. (2003). Tourism, peace, politics and ideology: impacts of the Mt. Gumgang tour project in the Korean Peninsula *Tourism Management, 24*(6), 675–685. doi:10.1016/S0261-5177(03)00047-5

Kim, S. S., Prideaux, B., & Prideaux, J. (2007). Using tourism to promote peace on the Korean Peninsula *Annals of Tourism Research, 34*(2), 291–309. doi:10.1016/0160-7383(90)90011-F

Lennon, J. J., & Foley, M. (2000). *Dark Tourism: The Atrraction of Death and Disaster.* Padstow, UK: Continuum International Publishing Group, Limited.

Leslie, D. (2012). *Responsible Tourism: Concepts, Theory and Practice.* Wallingford, UK: CABI

Lisle, D. (2007). Defending Voyeurism: Dark Tourism and the Problem of Global Security. In M. B. Peter & N. Marina (Eds.), *Tourism and Politics* (pp. 333–345). Oxford, UK: Elsevier.

Litvin, S. W. (1998). Tourism: the world's peace industry? *Journal of Travel Research, 37*(1), 63–66. doi:10.1177/004728759803700108

Litvin, S. W. (2000). Revisiting Tourism and Understanding. *Annals of Tourism Research, 27*(2), 526–529. doi:10.1016/S0160-7383(99)00090-0

Litvin, S. W. (2003). Tourism and understanding: The MBA Study Mission. *Annals of Tourism Research, 30*(1), 77–93. doi:http://dx.doi.org/10.1016/S0160-7383(02)00048-8

Lynch, P., & Causevic, S. (2008). *Tourism development and contested communities.* Retrieved 20 May, 2013, from http://www.espacestemps.net/en/articles/tourism-development-and-contested-communities-en/.

Macleod, D. V. L., & Carrier, J. G. (2010). *Tourism, Power and Culture: Anthropological Insights.* Bristol, UK: Channel View Publications.

Mannell, R. C., & Iso-Ahola, S. E. (1987). Psychological nature of leisure and tourism experience. *Annals of Tourism Research, 14*(3), 314–331. doi:10.1016/0160-7383(87)90105-8

Maoz, D. (2010). Warming up peace: An encounter between Egyptian hosts and Israeli

guests in Sinai. In O. Moufakkir & I. Kelly (Eds.), *Tourism, Progress and Peace.* Wallingford, Oxfordshire: Cabi International.

Marger, M. (2011). *Race and Ethnic Relations: American and Global Perspectives.* Belmont, CA, USA: Wadsworth.

McCabe, A. S. (2000). Tourism Motivation Process. *Annals of Tourism Research, 27*(4), 1049–1052. doi:10.1016/s0160-7383(99)00130-9

Minnaert, L., Causevic, S., Ateljevic, I., Morgan, N., Pritchard, A., & Pestek, A. (2013). *Critical Tourism Studies.* Retrieved 14 May 2013, from http://cts.som.surrey.ac.uk/.

Moufakkir, O., & Kelly, I. (2010). *Tourism, Progress and Peace.* Wallingford, Oxfordshire: Cabi International.

Nettekoven, L. (1979). Mechanism of intercultural interaction. In E. de Kadt (Ed.), *Tourism: Passport to development?* (pp. 135–145). Oxford, UK: Oxford University Press.

Novelli, M., Morgan, N., & Nibigira, C. (2012). Tourism in a post-conflict situation of fragility. *Annals of Tourism Research, 39*(3), 1446–1469. doi:10.1016/j.annals.2012.03.003

Noy, C. (2011). The Political Ends of Tourism: Voices and Narratives of Silwan/the City of David in East Jerusalem. In I. Ateljevic, N. Morgan, & a. A. Pritchard (Eds.), *The Critical Turn in Tourism Studies: Creating an Academy of Hope* (pp. 27–41). Amsterdam, The Netherlands: Elsevier Publications.

Nyaupane, G. P., Teye, V., & Paris, C. (2008). Innocents abroad: Attitude change toward hosts. *Annals of Tourism Research, 35*(3), 650–667. doi:10.1016/j.annals.2008.03.002doi:10.1016/j.annals.2010.03.006

Pearce, P. L. (1988). *The Ulysses factor: Evaluating visitors in tourist settings.* New York, NY: Springer-Verlag.

Pernecky, T. (2010). The Being of Tourism. *The Journal of Tourism and Peace Research, 1*(1), 1–22. doi:http://www.icptr.com/index.php/topics/issues/

Pizam, A., & Mansfeld, Y. (1996). *Tourism, crime and international security issues*. Chichester, UK: Wiley.

Plog, S. (1974). Why destination areas rise and fall in popularity. *The Cornell Hotel and Restaurant Administration Quarterly, 14*(4), 55–58.

Plog, S. (2001). Why Destination Areas Rise and Fall in Popularity. *Cornell Hotel and Restaurant Administration Quarterly, June.*

Poria, Y., & Ashworth, G. (2009). Heritage Tourism—Current Resource for Conflict. *Annals of Tourism Research, 36*(3), 522–525. doi:http://dx.doi.org/10.1016/j.annals.2009.03.003

Pritchard, A., Morgan, N., & Ateljevic, I. (2011). Hopeful Tourism: A New Transformative Perspective. *Annals of Tourism Research, 38*(3), 941–963. doi:10.1016/j.annals.2011.01.004

Putra, I. N. D., & Hitchcock, M. (2008). Terrorism and Tourism in Bali and Southeast Asia. In M. Hitchcock, V. T. King, & m. Parnwell (Eds.), *Tourism in Southeast Asia: Challenges and new directions* (pp. 83–98). Copenhagen, Denmark: NIAS – Nordic Institute of Asian Studies.

Redfoot, D. L. (1984). Tourist authenticity, touristic angst, and modern reality. *Qualitative Sociology, 7*(4), 291–309.

Report of the Secretary-General on the protection of civilians in armed conflict (2012). Retrieved September 5, 2012, from http://daccess-dds-ny.un.org/doc/UNDOC/GEN/N12/328/94/PDF/N1232894.pdf?OpenElement.

Rodgers, J. (2001). *AVCE Travel and Tourism Advanced*. Jordan Hill, Oxford, UK: Heinemann Educational Publishers.

Rojek, C. (1993). *Ways of Escape: Modern Transformations in Leisure and Travel.* Basingstoke, UK: Macmillan.

Salazar, N. B. (2006). Building a 'Culture of Peace' through Tourism: Reflexive and analytical notes and queries. *Universitas Humanistica, 62*(2), 319–333.

Scheyvens, R. (2011). *Tourism and Poverty*. London, UK: Routledge.

Scott, J. (2012). Tourism, civil society and peace in cyprus. *Annals of Tourism Research, 39*(4), 2114–2132. doi:http://dx.doi.org/10.1016/j.annals.2012.07.007

Seaton, A. V. (1996). Guided by the dark: from thanatopsis to thanatourism. *International Journal of Heritage Studies, 2*(4), 234–244. doi:10.1080/13527259608722178

Seaton, T. (2009). Purposeful Otherness: Approaches to the Management of Thanatourism. In R. Sharply & P. R. Stone (Eds.), *The darker side of travel: The theory and practice of Dark Tourism* (pp. 75–107). Bristol, UK: Channel View Publications, Limited.

Smith, M., MacLeod, N., & Robertson, M. H. (2010). *Key Concepts in Tourist Studies*. London, UK: SAGE Publications.

Smith, V. (2005). Tourism and Terrorism: The "New War". In J. Aramberri & R. Butler (Eds.), *Tourism Development: Issues For A Vulnerable Industry* (pp. 275–290). Clevedon, UK: Channel View Publications, Limited.

Spenceley, A. (2012). *Responsible Tourism: Critical Issues for Conservation and Development*. London, UK: Taylor & Francis.

Stephenson, M. L., & Ali, N. (2010). Tourism and Islamophobia: Muslims in Non-Muslim States. In N. Scott & J. Jafari (Eds.), *Tourism in the Muslim world* (pp. 235–252). Bingley, UK: Emerald Group Publishing Limited.

Tomljenovic, R. (2010). Tourism and intercultural understanding or contact hypothesis revisited. In O. Moufakkir & I. Kelly (Eds.), *Tourism, progress and peace* (pp. 17–34). Wallingford, UK: CABI.

Tornblom, K., & Kazemi, A. (2012). Advances in Justic Conflict Conceptualization: A new Integrative Framework. In E. Kals & J. Maes (Eds.), *Justice and Conflicts: Theoretical and Empirical Contributions* (pp. 21–52). Heidelberg, Germany: Springer-Verlag.

Tourism Concern. (2012). Ethical Tourists and Travellers. Retrieved September 17, 2012, from http://www.tourismconcern.org.uk/ethical-tourists.html.

Tourism4Peace. (2008). Egypt-Israel-Jordan-The Palestine Authority. Retrieved August, 21, 2012, from http://tourism4peace.net/background.html.

Tunbridge, J. E., & Ashworth, G. J. (1996). *Dissonant heritage: the management of the past as a resource in conflict*. Chichester: John Wiley & Son Ltd.

Uspomeneme 677. (2011). Retrieved 21 May 2013, from www.uspomene677.com.

Var, T., & Brayley, R. (1989). Tourism and world peace: Case of Turkey. *Annals of Tourism Research, 16*(3), 282–286. doi:10.1016/0160-7383(89)90078-9,

Var, T., Schluter, R., Ankomah, P., & Lee, T.-H. (1989). Tourism and World Peace: The Case of Argentina. *Annals of Tourism Research, 16*(3), 431–443. doi:10.1016/0160-7383(89)90055-8

White, L., & Frew, E. (2013). *Dark Tourism and Place Identity: Managing and Interpreting Dark Places*. Abingdon, Oxon, UK: Taylor & Francis.

World development report 2011: Conflict, security, and development. (2011). Washington, DC, USA: The World Bank. doi:10.1596/978-0-8213-8500-5

David Di Tota and Michael Lück

The Artificial Surf Reef Impacts at Mount Maunganui, New Zealand: A Study of Stakeholder Perceptions

1 Introduction

The search for the perfect wave is a common objective shared by most avid surfers around the world, yet various changing natural elements can cause a range of unpredictable wave sizes and activity proving it difficult to discover. In 1999, Kerry Black of ASR Ltd. set out to end this search for the perfect wave by creating one instead. His knowledge and experience with artificial reefs allowed him to develop a concept for an artificial reef that was designed solely for the purpose of creating consistent surfable waves. Surf tourism has been growing globally for years and the economic benefits of a surfable wave has only recently been realised. In an effort to harness the economic benefits of this growing global surf trend, Mount Maunganui at Tauranga, NZ was selected to host the first artificial surfing reef (ASR), due to its reputable surf and tourism culture.

The artificial surfing reef at Mount Maunganui was designed and built with the surfing community and tourism industry in mind, yet seven years after its completion the reviews and impacts of the reef on the community are still unclear. This chapter aims to uncover the impacts that the ASR has had on the surf culture and surf-related tourism outlets. A literature review and in depth interviews were conducted at Mount Maunganui to support the conclusions.

2 Artificial Reefs

Artificial reefs (AR) can be described as submerged structures (manmade or natural) that are deliberately placed on a seabed with the central purpose of mimicking the characteristics of a natural reef. They are intended to protect and encourage the development of marine ecosystems (Baine, 2001). For centuries, ARs have been used around the world as a popular resource enhancement technique which began in Japan in the 18th century. They were initially intended to support the fishing industry by providing various marine life with habitats that would promote protection and natural production. These early structures were made using natural materials, such as rocks or sunken tree branches but were later developed using manmade materials, including cinder blocks, sandbags, or scrap metal. While material made specifically for the purpose of creating artificial reefs are still employed today, other surplus materials are also used in some parts of the world, which include used tires, automobiles, train cars, aeroplanes, and sunken ships (Seaman & Lindberg, 2009). Stolk, Markwell and Jenkins (2007) reviewed research on artifical reefs as scuba diving resources, and proposed three main categories of artifical reefs: Simulation reefs, replication reefs, and transformation reefs (Table 1).

Table 1: Typologies of Artificial Reefs as Recreational Resources (TARRR)

Category	Appearance (underwater)	Typical Materials	Unit Complexity	Recreational Use Potential
(i) Simulation Reef (mimics form, aestethics and performance of natural reefs)	Natural	Cast limestone boulders Rock walls or reefs	Low	Fishing Diving Snorkelling Surfing Observation
(ii) Replication Reef (mimics performance of natural reefs, but not necessarily form or aesthethics)	Quasi-natural	Prefabricated modules Metal pyramids Concrete pipes Rubber tyres	Low – moderate	Fishing Diving Snorkelling Surfing Observation
(iii) Transformation Reef (mimics neither form, aesthethics nor performance of natural reefs, but produces unique conditions)	Non-natural	Piers and jetties Bridges Ships and planes Cars and carriages	Moderate – high	Fishing Diving Snorkelling Observation

Source: Stolk, Markwell, Jenkins (2007, p. 344)

Within recent years, a growing interest in artificial surfing reefs has emerged, spawning a new breed of multipurpose reefs. Aside from providing costal protection, ASRs are engineered primarily to improve wave quality and consistency for surfing. Artificial surfing reefs however are still a fairly new phenomenon that are within the early stages of their lifecycle, resulting in a lack of research and independent study regarding their environmental and socioeconomic impacts (Rendle & Esteves, 2011). Presently, seven ASRs have been installed at various beaches around the world but have had very little success. A number of projects have been abandoned or uninstalled due to poor performance or rising construction costs (Rendle & Esteves, 2011). Prior to the 1990s, the notion of constructing a reef solely for the purpose of generating a better surf had not been explored until a New Zealand based company, Amalgamate Solutions and Research Ltd., took on the challenge. ASR Ltd. is experienced in the design and construction of multipurpose artificial reefs and have installed a number of ARs around the world (ASR Ltd., 2010).

2.1 The Artificial Surfing Reef in Tauranga, New Zealand

By the mid 1990s, the first artificial surfing reef in New Zealand began to unfold as ASR Ltd. and the Artificial Reef Program (ARP) began to develop a proposal for Tay Street Beach at Mount Maunganui. Mount Maunganui is located on a peninsula in the city of Tauranga, within the Bay of Plenty region of New Zealand. It is situated on the east coast of the North Island, approximately 210km southeast of Auckland and is comprised of nearly 30,000 residents (Mount Maunganui Language Centre, 2009). Due to the peninsula's unique shape (Figure 1), tourists and residents of Mount Maunganui enjoy a calm and sheltered bay on the inner harbour side, while also being able to enjoy the numerous waves and white sand beaches on the ocean side.

Figure 1: Mount Maunganui and location of the artificial surfing reef (Source: M. Lück)

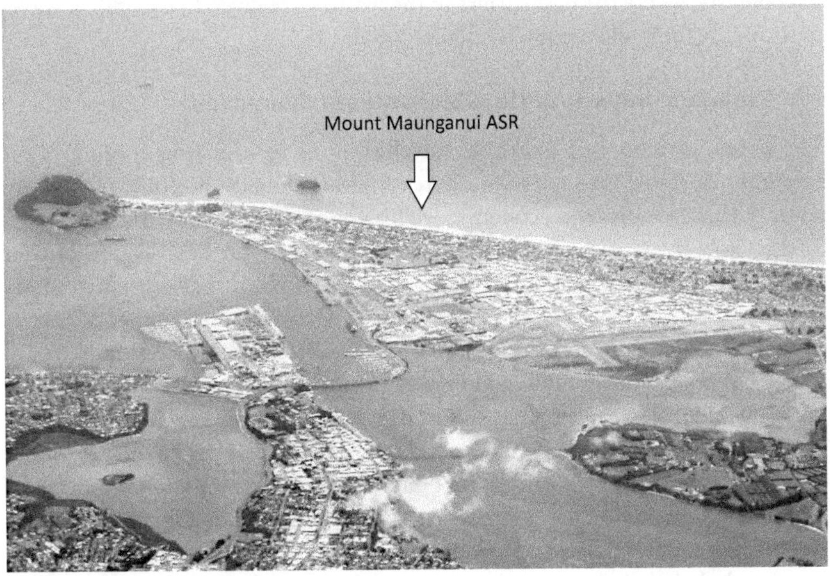

The town of Mount Maunganui (locally referred to as 'The Mount') is named after the extinct volcano that rests at the northern tip of the peninsula, which

Māori call 'Mauao'. Surfing, swimming, fishing, and shopping are among the top activities enjoyed at the Mount throughout the year (Tourism New Zealand, 2011). Mount Maunganui, specifically Tay Street beach, was chosen as the optimal location for the first artificial surfing reef based on a point scoring system that considered several factors which included, access to the University of Waikato (for research purposes), sourcing of construction materials, perceived erosion problems, conservation significance, and the local surf culture (Mead & Black, 1999). Due to his heavy involvement with the University of Waikato, Black decided to develop New Zealand's first ASR in partnership with the university's research program to ensure future enhancement of further development opportunities. Mead and Black (1999) stated that surfing had increased in popularity worldwide over the past 30 years and demand for quality waves was stronger than ever. Estimates about the number of surfers range from 10 million (Buckley, 2002) to 23 million surf-riding tourists globally (Orams & Towner, 2013). The Mount Beach ASR ('Mount Reef') was designed and built to take advantage of this trend by producing a consistent and 'perfect wave' for optimal surfability. While the reef was still under construction, Black mentioned in an interview with Surfingmagazine.com that he imagined being able to host surfing contests, where the best surfers in the world were given five identical waves to showcase their talent equally (Black, 2004).

2.2 Economic Impacts of the ASR in Mount Maunganui

The potential economic worth of surfable waves to a surfing community is immense. Surfable waves are considered highly valuable and finite natural resources that can provide economic well-being if utilised properly (Peryman, 2011). Globally, it is estimated that surf-riding tourism has a value of over US$8 billion (Carroll, 2004, in Orams & Towner, 2013). The feasibility study conducted for the Mount ASR project stated that, given the number of surfers the beach could support at one time and considering the spending habits of surfers, the local economy could expect to earn NZ$1.6 million annually, or NZ$500,000 if the ideal surfing conditions occur one third of the time (Mead, 2009). Economic benefits would include revenue growth from such activities as food and beverage purchases, fuel purchases, accommodation rentals, surf equipment rentals/purchases, as well as expenditure for other local tourism activities (Scarfe, Healy, Rennie & Mead, 2009). However, contradicting research has concluded that there has been far too little academic investigation into the value of recreational surfing and its impact on both residents and tourists in specific locations (Lazarow, Miller & Blackwell, 2009). While the project's feasibility study conducted by Mead determined the Mount Reef's economic value at approximately NZ$500,000 annually, other surf communities around the world have valued their own surf breaks anywhere from US$300,000 to US$400 million, using various methods of evaluation (Lazarow, Miller & Blackwell, 2009). Some argue that because the artificial surfing reef industry is dominated primarily by two major companies (ASR Ltd. and International Coastal Management), there has not been any attempts to provide a framework for the investigation of the market value of recreational surfing, with the specific intention of using the vague data to their own advantage (Lazarow, 2007).

There are however several other direct, indirect, and passive economic benefits that together form a total economic value for the Mount Reef. For example, a direct value would include the initial intentions of the reef, such as producing a consistent wave that would increase surf tourism. An indirect value would include any benefits that the reef provides that would help support the direct values, such as coastal protection or marine life production/protection. Finally, a passive value that would contribute to the total economic value of the ASR is the overall knowledge that the objective was reached and the efforts put forth to construct the reef has resulted in increased tourism (Whitmarsh, Santos, Ramos & Monteiro, 2008). Although the monetary valuation of some of these effects can be determined, it does exceed the scope of this study, therefore a simple positive or negative impact assessment of these economic values will be determined.

2.3 Socio-cultural Impacts of the ASR in Mount Maunganui

Irrespective of the positive economic gains that would accompany a prosperous surf break or ASR, socio-cultural consequences resulting from the degradation of a healthy surf amenity such as an ASR must also be considered. Should a surf community suffer from the loss of a surfable reef, it could face a "decrease in trust in government and loss of local sovereignty; increased negative social impacts on other already crowded surf breaks; people may turn away from surfing; and/or negative impacts on local and visitor perceptions" (Lazarow, Miller & Blackwell, 2009, p. 153). None of these risks seemed to have been assessed in the original Mount Reef proposal or feasibility study.

A unique feature that was included in the proposal was the ability to remove the reef five years after construction had completed, should the performance of the reef be less than par (Mead & Black, 1999). At the time of this study, it had been nearly seven years since the completion of the Mount Reef, yet despite the project exceeding its initial budget, taking longer than expected to complete, and the negative effects it has displayed, it has yet to be removed. Many have stated that the reef is producing dangerous and inconsistent waves and it would not be missed if it were to be removed (Bay of Plenty Times, 2011). Black blames the poor performance on dire weather conditions and construction veering from the initial designs, but believes the reef is still a valuable research site. Similar cases have occurred in Opunake, New Zealand and Bournemouth, UK where over-budget ASRs were installed. They promised economic growth, that ultimately resulted in dangerous waves (Keith, 2011) and impossible to surf conditions (The Telegraph, 2010).

Whether considered a major surf attraction or an expensive flop, the ASR in Mount Maunganui has generated a lot of contradicting reviews. Tourism New Zealand's website advertises Mount Beach as a "surfing mecca" that "produces consistently perfect waves", and that it is "suitable for competent and experienced surfers" (Tourism New Zealand, 2011). However, a number of news articles and New Zealand surfer blogs have stated the opposite. Many avid surfers have even admitted that they never even surfed the beach due to its poor and inconsistent wave conditions (Ramboestrada, 2009). Clearly, there is some confusion as to how the reef has actually affected the surfing culture and tourism in Mount Maunganui over the years, and this study aims to identify some of the key impacts.

3. Methodology

Semi-structured interviews were conducted with various stakeholders in Mount Maunganui and Tauranga. Respondents were called ahead of time to set up appropriate times and locations to meet. A number of respondents offered various and unique perspectives, in that they represented diverse demographics at the Mount. Table 2 illustrates where each of the respondents were met and provides a brief overview of the knowledgebase/perspective they offered.

Table 2: Interview Location and Respondent Knowledgebase

Respondent	Interview Location	Knowledgebase/Perspective
A	Mount Maunganui Lifeguard Office	Office administrator; Grew up in Mount Maunganui; Local perspective
B	Surf Shop A	Retail shop owner/operator
C	Surf Shop B	Shop employee; Surf enthusiast
D	Respondent's home	Professionally represented NZ in a number of surf competitions; Lifeguard; Surf School owner/operator; NZ surf website administrator; Lives and operates at Tay Street Beach

Respondent	Interview Location	Knowledgebase/Perspective
E	Mount Beach	Surf school/retail shop owner/operator; Heavily involved in Mount ASR development
F	Respondent's home	Surf school owner/operator; Foreign perspective; Moved to NZ from England in the early 1990's
G	Respondent's home	Surf school owners/operator; Foreign and local perspective; Husband from NZ; Wife from USA; Business partners
H	Accommodation	Hostel/lodge owner/operator

Interviews with managers, owners, employees, participants, and/or volunteers of surf-tourism related businesses took place on May 4th and May 5th, 2012. Observations were also noted regarding tourist activity on Mount Maunganui's beaches during this timeframe.

3.1 Limitations

A number of limitations were noted that may have affected the outcome of this study. These limitations are as follows:

- Possible negative effects on tourism that may have arose from the cargo ship *Rena* oil spill off the Astrolabe Reef in Tauranga in early October 2011.
- Although several of the owners/operators had been residing in Mount Maunganui prior to the installation of the ASR, a number of their businesses were not operating at the Mount until after the ASR was installed.
- The weather conditions at Mount Maunganui during the research period produced little to no swells/waves at the Mount Beach, making it difficult to observe regular surf activity at the Mount ASR.
- A number of stakeholders have declined interview requests, each claiming that they have no association or comments with regards to the surf culture or the Artificial Surfing Reef at the Mount.

4. Results and Discussion

In addition to the interviews, observations indicated that the Mount bears a very active tourist lifestyle. Despite the lack of waves for surfers to capitalise on, the beaches, cafés, and shops seemed very busy. Tourists and locals participated in a number of other water-based activities including stand-up paddle boarding and kite boarding, both of which do not require any waves. The section of beach perpendicular to the ASR did seem very sparse over the weekend, however it occupies such a small portion of the large beach, that it almost goes unnoticed.

4.1 The Organisations and Mount Maunganui

The number of years each respondent's business or establishment had been operating at the Mount was asked in order to get a better understanding of the organisation's age in relation to the ASR (Figure 2).

Figure 2: Years in operation, and establishment of ASR

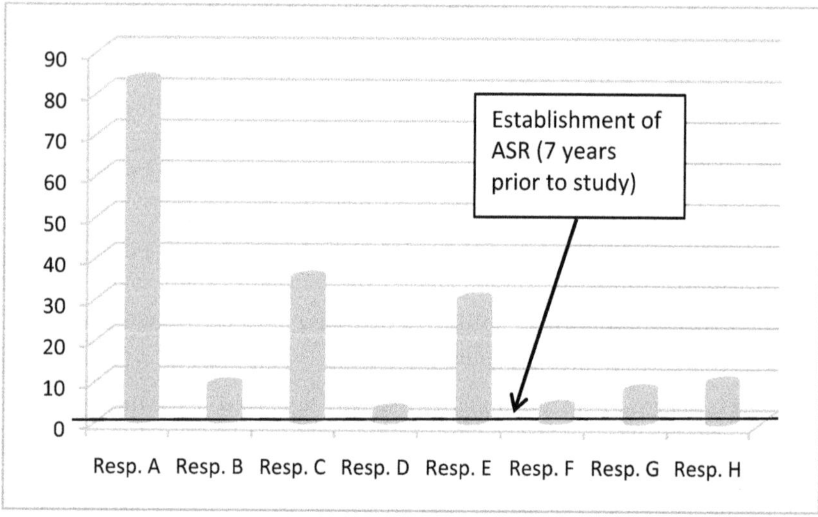

Only three of the eight respondents interviewed were employed at an institution that had been established prior to the ASR being installed. Respondent A's establishment had been operating at the Mount Beach for over 80 years as the beach's predominant lifeguard service and club. Most of the respondents started their business venture either after the ASR had already been installed or while it was still being developed. Considering the majority of businesses are still relatively young, opportunities must have been perceived available at the Mount, even a number of years after the reef had been constructed.

Respondents were asked to describe their target markets, and those of Mount Maunganui in general. The question also helped determine the respondents' overall view of the surf culture at the Mount. The responses illustrate the type of tourists that frequent the Mount and its present social environment (Table 3).

Table 3: Target Markets and Current Patronage at Mount Maunganui

Respondent	Target markets and current patronage
A	Good mix of international, domestic tourists, and local people.
B	Anywhere from 2 to 60 years old. Mainly families and surf enthusiasts. Good mix, but mostly locals and backpackers through the summer.
C	A lot of locals during the winter, and about 90% tourists in the summer.
D	Very broad. Domestic tourists and families in middle of summer, and backpackers during the shoulder seasons. Kids to older people.
E	Mainly local (70–85%), but some tourists over the summer – all ages.
F	More locals and regulars, occasional tourists (mostly in summer), but mainly women (14–30 years old).
G	All ages, a lot of Europeans (tour groups). About 60% Kiwis, 40% tourists.
H	A lot of Europeans, anywhere from 16–50 years.

Based on the responses, it appears that many locals utilise the services provided at the Mount, despite the fact that the Mount is visited by locals and visitors equally. The summer months seem to attract more backpackers, Europeans, and other tourists, while shoulder seasons draw more families and locals of all ages.

4.2 Surf Culture and Surfers' Skill Levels at Mount Maunganui

Respondents were asked to describe surfing and the surf culture at the Mount. The majority of respondents indicated that the Mount is one of the most popular surf beaches, with the biggest population of surfers per capita in New Zealand. The culture was described as strong, for example, "tight bunch, everyone knows everyone. Good culture" and "very strong. Not just surf culture, but board riding in general. Kite boarding, stand-up paddle boarding, body boarding. More opportunities to catch waves." While one respondent noted that the surf culture "can be competitive", another felt that it is "very friendly, not territorial."

Overall, the perception of the surf culture at the Mount is very positive, considering the strong and friendly surf community.

Orams and Towner (2013) proposed a model, describing various typologies of surfers, based on the difficulty of the location, and the skill levels of the surfers (Figure 3).

Figure 3: The Orams and Towner typology of surf-riding (Orams & Towner, 2013, p. 184)

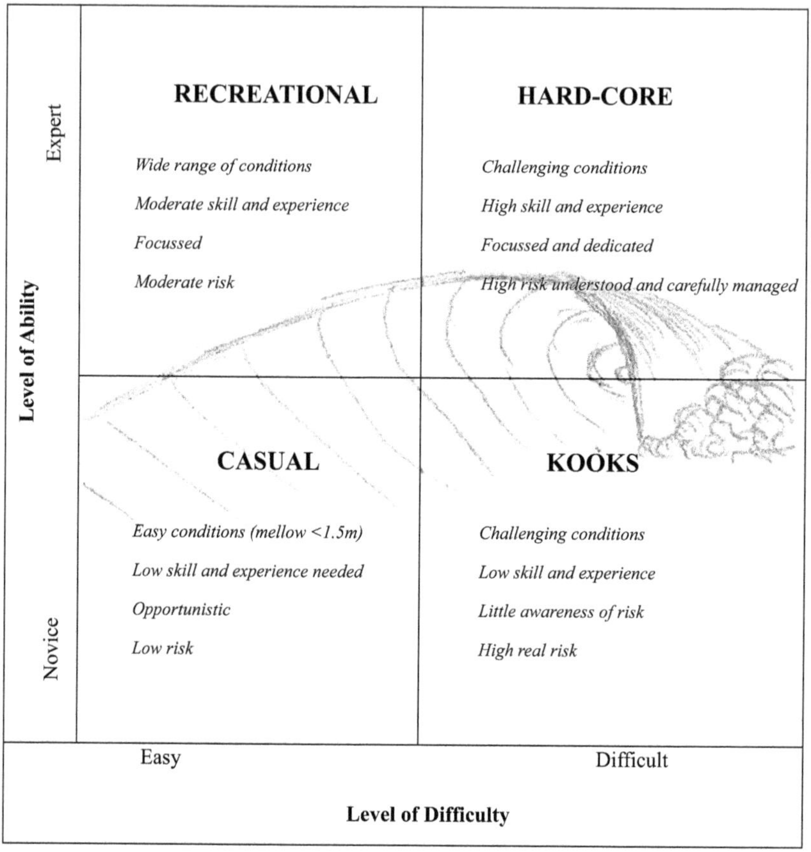

When asked what skill levels most surfers are that visit the Mount, respondents noted that "everybody", "all ranges", and "beginners all the way to experts" are frequenting the area. Thus, based on the responses in this study, all levels of surfers can be found in the Mount. One respondent noted that the skill level is related to the surf on the particular day: "if [the surf] is really good, you get more advanced surfers coming in." Subsequently, it appears that the surf skill required to enjoy the Mount varies depending on weather conditions and location of surf, but can generally be enjoyed by surfers of all skill levels. While the Mount accommodates surfers of all skill levels, the surf itself is not overly challenging, and thus surfers at the Mount would fall mostly into the "casual" and "recreational" categories of Orams and Towner's typology.

4.3 The Artificial Surf Reef (ASR) at Mount Maunganui

Each of the respondent's general knowledge of the Mount ASR was sought to determine the level of awareness and comprehension of the presented topic. The replies helped verify the personal impacts felt by each respondent, both before the ASR was installed and after (Tables 4 and 5).

Table 4: General Knowledge about the ASR

Respondent	Response
A	Initially designed to help stop dune erosion, then became a surf reef.
B	Surf has never been as good since the reef was installed. Good breaks on either side though. Not what it once was. Never finished. An expensive test run.
C	It's crap. Put a lot of personal time into it. Packed sandbags, raised money through buying and selling 'Mount Reef' t-shirts. Was supposed to be amazing. A lot of the surfing community did fund raising to help with the project. Voluntary work. Had great potential and started to work, but swells ended up wrecking it before it was completed. Wasn't done properly. Now about 500m of beach is useless for surfers.
D	Was built to improve the quality of the waves, but hasn't worked at all.
E	Was involved with the project from the start. Helped form the Reef Trust to get the concept accepted and apply for resource consents. Raised money and put the project together to get constructed. Personally invested a huge amount of money and about 13 years of time and effort.
F	Doesn't work. It's not finished.
G	It was a failed project, not finished, designed for competitive surfing.
H	Was a good idea, but doesn't work. Personally invested a lot of money into the Reef Trust and lost it.

Table 5: Initial Reaction to the ASR, and Feelings Afterwards

Respondent	Response
A	Took a lot of money from various grants that could have gone to better things.
B	Disappoints me to hear that they continued to install reefs around the world, despite the failure. We sold t-shirts to raise money for the community. We felt obligated, being a big part of the industry, but were not deliberately "pro-reef builders".
C	Excited at first. Waste of space and want them to pull it out now. Would probably be better waves.

Respondent	Response
D	I was excited at first. I worked for people who invested money but I moved away while they were constructing it. ASR Ltd. had falsely advertised the reef through fake photos of good surf, claiming it as "*The* Reef", but it's not. Tried to marginalize any negative press. Very upsetting.
E	Very excited in the infancy stage, that's why we got involved with the idea. Brought a lot of competitive businesses together to contribute in one way or another, and I think it brought the greater majority of the surfing community together, but now I think it's done the opposite because people have taken a very negative attitude towards the results. It's disappointing to hear people talk about it and it's difficult from my point of view mainly because they are just so uninformed, even though they make out that they know everything about it.
F	I thought it sounded awesome. Used to be a great place to surf, but not as popular anymore.
H	I thought it was a good idea and bought into it just like everyone else. It was a real let down.

In general, the Artificial Surfing Reef has instilled a very negative impression in the minds of nearly all the respondents. Every respondent seemed very optimistic about the possibilities of the ASR during the early stages of the project, but were ultimately let down when the reef did not deliver what was promised. It is unanimously agreed upon by all respondents that the reef does not work and it has damaged the surf break that once was there prior to the construction of the ASR. These findings suggest that the direct and passive economic values regarding the function and affirmative knowledge of the reef's functions are in fact negative and may devalue the reef's total economic value. Not only did the ASR not produce the waves that is was supposed to, but it also harmed the break that was previously at that location.

4.4 The Impact of the ASR on Local Businesses

Respondents were asked whether they have noticed any change to their business levels or the tourism market at Mount Maunganui, in relation to the installation of the ASR (Table 6). Generally, the respondents' opinions regarding the reef's ability to increase tourism in Mount Maunganui varied. A number of respondents claim that it has not, while others believe that it has drawn tourist interest, despite its failure to produce the waves it should have. Given his heavy involvement with the project, one respondent has remained optimistic about the reef's potential. He explains that the construction of the reef did not go according to plan, but can still be revived if the proper resources were available. However, after the community has already contributed to the cause and many have suffered personal losses, they were left with a cynical attitude towards the ASR. Some respondents found it difficult to attribute an increase in tourism directly to the ASR, or to a general upward trend in tourism.

Table 6: Did the ASR succeed, and did it generate tourism to Mount Maunganui?

Respondent	Response
A	Failure.
B	No. The uneducated may come because of what they heard, and are disappointed at the perception of a reef, but it hasn't hurt tourism.
C	Not at all. It's a joke now. You can't surf the reef at all. Didn't damage tourism, but didn't help either. There would be thousands of people here if it worked, but they did it wrong.
D	Not at all. The interest in the reef is more in the people who don't really surf, but want to know where it is and what's going on. Nowhere near what they wanted to achieve. Interest in surf tourism has shifted to Matakana Island due to overexposure over the last 3 to 4 years, and it's natural. The ASR is well publicised and tourists do come looking for it, more out of interest. Tourism numbers have increased since the installation of the reef, but it's not because of the reef. The tourism infrastructure at Mount Maunganui has grown and drawn people in. Numbers have dropped off though since the Rena incident* (half the numbers this summer).
E	Increased interest and focus on surf tourism in the beginning, but not so much anymore. People are interested in the reef, but not so much for surfing. The reason it doesn't work is simply because it's incomplete. It's about 50–60% finished because the contractors never tied the bags off like they should have. The foundation is laid and there is still great potential, but more funding is required, but the resources are exhausted.
F	People would come to see it just because of what they've seen or heard. It has increased interest, but very rarely.
G	I think a little bit, a lot of people ask about it because it's listed in the Lonely Planet travel guide.
H	There hasn't been any change in the tourism or business trends. It was busy during the construction period, but it slowly died off after a couple years.

*The respondent refers to the container ship *MV Rena*, which ran aground on the Astrolabe Reef off the coast of Mount Maunganui/Tauranga, on the 5th October 2011. The incident is widely seen as the worst environmental maritime disaster in New Zealand's history.

When it comes to their own businesses, the majority of respondents did not notice any change. Only one respondent noticed a decrease, but was not sure whether this is because of the ASR, or because of a general recession. Some respondents felt that it created an initial increase in visitors, mostly out of curiosity, but that it has dropped since, and is not attracting repeat visitors: "May have increased due to people coming to check it out initially to see if it works. Doesn't attract repeat tourists.", and "it did create foot traffic in the beginning for my retail shop, but not anymore. A lot of people don't even talk about it anymore."

Although it has not necessarily increased surf tourism as intended, the reef has allowed for other business opportunities. A surf school operator mentioned that he will be expanding his business ventures to include snorkelling options that explore the marine life that is a by-product of the reef: "No [change] for the surf school, but I am definitely looking in to expanding my business into snorkelling on quiet days. It's not the best location, but it does have potential for snorkelling families with good conditions." Another respondent has also recently expanded his business ventures to include other activities that do not rely on waves, such as stand-up paddle boarding and kite-boarding. It was observed during the research period, that when compared to the other surf schools that were interviewed, and despite the lack of waves at the Mount, that respondent was the only surf school that was busy with customers. This suggests that the reef has produced a positive indirect economic value in its ability to produce a desirable marine habitat and growth into other surf outlet possibilities.

4.5 Safety at the Mount Maunganui ASR

One question was specifically directed at the two respondents with a heavy involvement in rescue services at the Mount, with the objective to determine whether the ASR has posed a greater public threat since being installed. While one respondent indicated that there has been no change ("was a problem area since before the reef"), the second respondent noted that the ASR "created more problems in that area due to rips. Public facilities (bathrooms and cafés) installed in the same area of beach where the reef is located draws people to the more dangerous area and causes problems. More problems due to higher numbers of people." In addition, another respondent mentioned that he does not teach his clients how to surf near the reef because he believes the rips are too dangerous, due to the rip currents created by the reef.

These findings do support the Bay of Plenty Times (2011) claims that the reef produces dangerous and inconsistent waves; however it is still uncertain as to whether the reef itself can be completely held accountable for that. As the second respondent mentioned, the higher number of incidents that occur in the area may have resulted due to the increased public facilities (i.e. toilets) that have emerged in that section over the years. Tay Street (off the beach where the ASR is located) is a popular location for surf events in Mount Maunganui, and the facilities are required to serve the spectators. Therefore, the public is naturally drawn to this potentially hazardous part of the beach to use the facilities, even when no events are being held.

4.6 Surfing Events at the Mount Maunganui ASR

Given the popularity of the beach for surfing events, respondents were asked whether the ASR has increased the number of events, as suggested in the initial proposal for the construction of the ASR. The unanimous response was that the ASR has not resulted in an increase of surfing events. In fact, the majority stated that the location of the beach is good for spectators, but the actual surfing during the events does not take place on the ASR, but on either side of it: "There has been competitions but not at the reef, about 100m north. They're still held at the same beach, but the reef is avoided."; "Never on the reef, but events are held at the beach all the time. It's too far out, and good for snorkelling, paddle boarding, and shell fishing, but that's not what it was built for."; and "the location is great for spectators given the vantage point and drive-by traffic, but the reef isn't used directly, but they used the breaks on either side of it." One respondent noticed an increase of people involved in the events though, stating that "there are more people involved in competitions than there was before, but I think that's mainly because of better advertising." Thus, surf competitions are still held at the Mount beach given its ability to host large crowds. The reef however is not utilised and is actually avoided altogether as it creates a large obstacle. This is supported by other surfers that blog about the ASR producing inconsistent wave conditions (Ramboestrada, 2009).

One would imagine that each competition held within the vicinity of the reef would produce a subtle reminder of the failed project and would conjure up negative emotions amongst the local community. These findings also contradict Black's (2004) intentions to design a consistent wave that would host surfing contests, by actually creating a reef that would have the opposite effect.

4.7 Other Comments on the Mount Maunganui ASR

Finally, respondents given the opportunity to divulge any additional information about the ASR that had not already been discussed. It was to allow an open window for general discussion or concerns that the respondents may have felt were over-looked by the other questions (Table 7).

Table 7: Additional Comments Regarding the Mount Maunganui ASR

Respondent	Response
A	Didn't deliver what it promised, same thing happened in England. Hasn't hurt or helped tourism.
B	I don't like seeing them selling their skills to other places based on this failure. Can they take it out? Expensive research site. Cost was blown out of proportion, it was ridiculous.
C	They should pull it out, but they can't now because of the marine life that is living there now. You can stand on the reef during low tide. It had potential, but they didn't do it right. The reef built by ASR in Australia worked because they filled the bags first then dumped them, but at the mount they sunk the bags and then tried to fill them. If they did it all at once, it may have worked, but they did it in sections instead.
D	They positioned it too far and for big swells, instead at looking at more majority conditions and building it off that. If they tried to build something that would help enhance small nice days, it would have been useful for the surf schools and local surfers. Still a great city to visit for surfers with good waves. Great destination for families.
E	We have a strong surf culture at the Mount. If funding was there, it could work but who knows if that'll be achieved.
F	I think they should just build a pier over the reef like at Huntington Beach in California, which would create waves and add an attraction for people to walk out on to.
H	Confuses tourists more than anything. People come expecting one thing and are disappointed when they actually find out how it turned out.

When given the opportunity to speak freely about the ASR, most respondents replied with negative remarks regarding its effect, however several followed up with positive and reinforcing comments about the Mount's surf culture.

5. Conclusions and Recommendations

After the creation of an artificial surf reef at Mount Maunganui, New Zealand, the ASR has been widely criticised for not providing the promised "consistent good surf wave" and a resulting increase in tourist numbers, especially for surfing events. This study employed a qualitative approach by interviewing a number of stakeholders with an interest in the ASR, ranging from business owners to surf live saving club members.

Based on these interviews, it can be concluded that although the Mount reef has not worked out as it was initially intended to, the surf culture and tourism industry in Mount Maunganui remained strong. Promises of increasing the levels of tourism at the Mount may not have been fulfilled, yet it has not been damaged either. If the ASR does not work as it was intended to, then these findings challenge the belief that a community that suffers from a lost surfable reef would face "negative impacts on local and visitor perceptions" (Lazarow, Miller & Blackwell, 2009, p. 153). Mount Maunganui still offers its locals and tourists an active lifestyle through a number of outlets and various surf locations aside from the Mount ASR. Mount Maunganui still hosts a strong and reputable surfing community in New Zealand given the array of prime surfable locations. The absence of a small stretch of beach that was considered hazardous prior to the reef being installed has not necessarily deterred the surfing community, but simply irritated it through disappointment. Surf tourism may not have increased at the Mount due to the ASR; however a number of other business opportunities have been realised in light of its existence, such as snorkelling. A shift from traditional surfing to new activities has also been recognised, such as kite-boarding and stand-up paddle-boarding, and is offering an alternative business solution to surf related organisations at the Mount. It is recommended that these surf related businesses continue to focus on both traditional surfing activities, as well as alternative water-based activities that do not rely on waves. Becoming less reliant on sporadic wave conditions will ensure that the industry remains resourceful and covers a larger portion of the adventure-seeking tourism market.

Although some believe the ASR should be removed from the beach altogether, it is recommended that the reef remain at Mount beach, as it still provides some positive by-products which include a valuable marine habitat that doubles as a snorkel/dive location, as well as a practical research location for studying marine biology. Some respondents also believe that the ASR still has potential to provide the perfect and consistent wave that it was originally intended to, however until the community's wounds have healed and the additional resources can be provided, that potential may never be realised, and the search for the perfect wave will continue.

Final Note

During the process of writing and the final revisions of this chapter, it was reported that the Bay of Plenty Regional Council has decided to remove the reef, due to the risk it poses for swimmers and surfers. The council deputy chief executive is quoted as noting that the reef "also generated some unforeseen effects, including creating a large scour hole which affects waves and currents, increasing frequency and intensity of rips which pose a serious risk to swimmers" (Morton, 2014).

References

ASR Ltd. (2010). *Who We Are.* Retrieved from http://www.asrltd.com/who-we-are/index.php.

Baine M. (2001). Artificial reefs: a review of their design, application, management and performance. *Coastal Management*, *44*(3–4), p. 241–25.

Bay of Plenty Times. (2011, April 7). *Surf reef branded a dangerous flop.* Retrieved from http://www.bayofplentytimes.co.nz/news/surf-reef-branded-a-dangerous-flop/1050916/.

Black, K. (2004, June 7). Interview by M Walker [Web Based Recording]. 60 seconds. *Surfer Magazine*, New Zealand., Retrieved from http://www.surfing-magazine.com/magazine/jul04_60seconds/.

Black, K., & Mead, S. (2009). Design of surfing reefs. *Reef Journal*, *1*(1), p. 177–199.

Buckley R. (2002). Surf tourism and sustainable development in Indo-Pacific Islands: The industry and the islands. *Journal of Sustainable Tourism*, 10, 405–424.

Keith, L. (2011, March 23). *Surf reef trust left high and dry.* Retrieved from http://www.stuff.co.nz/taranaki-daily-news/news/4798348/Surf-reef-trust-left-high-and-dry.

Lazarow, N. (2007). The value of coastal recreational resources: a case study approach to examine the value of recreational surfing to specific locales. *Journal of Coastal Research*, 50(50), 12–20.

Lazarow, N., Miller, M. L., & Blackwell, B. (2009). The value of recreational surfing to society. *Tourism in Marine Environments*, *5*(2–3), 145–158.

Mead S. and Black K. (1999). A multipurpose, artificial reef at Mount Maunganui Beach, New Zealand. *Coastal Management*, *27*(4), 355–365.

Mead S. (2009). Multiple-use options for coastal structures: Unifying amenity, coastal protection and marine ecology. *Reef Journal*, 1, p. 291–311.

Morton, J. (2014, April 16). *NZ's first artificial surf reef wiped out.* New Zealand Herald, retrieved from http://www.nzherald.co.nz/nz/news/article.cfm?c_id=1&objectid=11239213.

Mount Maunganui Language Centre. (2009). *Mount Maunganui Town.* Retrieved from http://www.study-english.com/about.php?The-Town-2.

Orams, M.B. & Towner, N. (2013). Riding the wave: History, definitions, and a proposed typology of surf-riding tourism. *Tourism in Marine Environments,* 8(4), 173–188.

Peryman, B. (2011). *Bay of plenty surf break study.* Lincoln University. Faculty of Environment, Society and Design. Department of Environmental Management.

Ramboestrada. (2009, June 11). *The truth about the mount reef.* Retrieved from http://ramboestrada.com/2009/06/11/the-truth-about-the-mount-reef/.

Rendle, E. & Esteves, L. (2011, May 9). *Developing protocols for assessing the performance of artificial surfing reefs -a new breed of costal engineering.* Littoral 2010 – Adapting to Global Change at the Coast: Leadership, Innovation, and Investment 2011.

Scarfe, B. E., Healy, T. R., Rennie, H. G., & Mead, S. T. (2009). Sustainable management of surfing breaks – an overview. *Reef Journal, 1*(1), p. 44–73.

Seaman, W., & Lindberg, W. J. (2009). *Artificial reefs.* In: J. H. Steele et al. (eds.) Encyclopedia of Ocean Sciences (second edition), pp. 226–233. Amsterdam, Elsevier.

Stolk, P., Markwell, K., & Jenkins, J. M. (2007). Artifical reefs as recreational scuba diving resources: A critical review of research. *Journal of Sustainable Tourism, 15*(4), 331–350. doi:10.2167/jost651.0

The Telegraph. (2010, March 05). *Bournemouth artificial reef 'near impossible' to surf.* Retrieved from http://www.telegraph.co.uk/news/newstopics/howaboutthat/7374343.

Tourism New Zealand. (2011). *Mount Maunganui Beach, Bay of Plenty – New Zealand.* Retrieved from http://www.newzealand.com/int/article/mount-maunganui-beach/.

Whitmarsh, D., Santos, M. N., Ramos, J., & Monteiro, C. C. (2008). Marine habitat modification through artificial reefs off the algarve (southern portugal): An economic analysis of the fisheries and the prospects for management. *Ocean & Coastal Management,* 51(6), p. 469–468.

John S. Hull

The French Shore Historical Society: Creating New Tourism Experiences for Visitors

1. Introduction

Cultural tourism is increasingly recognized as an important motivation for international travelers in search of experiencing the cultural heritage, arts, philosophy and institutions of another region or country (Timothy & Boyd 2003). The UNWTO reports that 37% of all international trips include a cultural component and that cultural tourism is growing at 15% annually (UNWTO 2011). In Europe, the Association for Tourism and Leisure Education (ATLAS) Cultural Tourism Survey (2007) ascertains that the proportion of tourists on cultural holiday has increased from 17% in 1997 to 31% in 2006. In North America, the Travel Activities and Motivation Survey of Canadian Heritage Enthusiasts (TAMS) (2003) determined that 32% of cultural travelers add extra time to their trip, because of cultural, art, historic or heritage experiences.

The global interest in culture has resulted in the development of new niche markets linked to cultural and human heritage. In the 1970s cultural tourism became recognized as a distinct product category, pursued by a small number of educated, affluent tourists (McKercher and du Cros 2002). As global demand for cultural experiences have grown, local, national and international bodies are increasingly recognizing the role of cultural tourism in preserving world heritage, supporting national identity, and as a vital means of economic support. As a result, cultural tourism is evolving to become a mass form of tourism that offers a broad range of heritage, popular culture and living cultural attractions that integrate the "everyday life of local communities" (Richards 2007, 2).

2. Defining Creative Tourism

Richards and Wilson (2006) argue that creative tourism is one new form of cultural tourism. While cultural tourism is considered a passive observation of things of the past such as artifacts, creative tourism is the transfer of the past into the present and the future via communication between locals and visitors (Table 1). The main assets of creative tourism are manifested through local cultural traditions and local communities. These assets are organized into visitor attractions and experiences that promote active skill development and participatory opportunities between hosts and guests. As a result, local inhabitants are motivated to share their history and culture with visiting tourists preserving the originality of a place.

Table 1: Cultural vs. Creative Tourism[1]

Type of tourism	Time	Cultural Focus	Consumption	Learning
Cultural tourism	Past and present	High culture, popular culture	Product, Process	Passive
Creative tourism	Past, present, future	Creative process	Experience, co-production between tourist and local	Interactive, Active skill development

This growth in creative tourism represents a shift in the tourism sector away from a focus on the tangible resources such as the built and material heritage to more intangible resources focused on the local traditions. This movement is creating many challenges and opportunities for destinations who have had to focus on their uniqueness to 'create' experiences that attract visitors (Richards & Wilson 2006).

Recent changes in market demand indicate that there are growing numbers of cultural travelers who are going on holidays in search of experiences that offer opportunities for personal- and self-development. The shift from tangible to creative intangible attractions is mainly driven by a new creative class (Florida 2002). Florida argues that the rise of the creative worker signals the rise of the creative traveler. The creative class is knowledge-based, labor-intensive, and paid for creating something new (performing arts, crafts, film, sport, publishing, architecture, software). This creative tourist is traveling to engage in new experiences that are an extension of everyday life and work. They are looking for destinations that offer the option of self-development (Florida 2002).

The purpose of this article is to first identify the strategic development of creative tourism experiences. Second the application of these strategies will be analyzed in a context of a cultural case study from the French Shore in Newfoundland and Labrador, Canada where a local historical society is promoting and marketing the unique cultural heritage of the region. The last section will analyze how creative tourism is serving as a strategy for attracting tourists and overcoming challenges facing tourism destinations.

---

1 Richards, G./Wilson, J. (2006). *Developing Creativity in Tourist Experiences: A Solution to the Serial Reproduction of Culture?* Tourism Management 27.

3. Strategic Development of Creative Tourism Experiences

In response to the growing market demand, policymakers globally are proposing a number of strategies to foster the development of creative experiences as part of the new experience economy (Pine & Gilmore 1999). Based on the book "*Creative tourism – a global conversation*" Wurzburger, Aageson, Pratt, & Pattakos (2009) argue that the following five steps need to be considered to support creative tourism development at a destination:

- Engage visitors in constructing the experience of a destination.
- Develop public programming: courses, conferences, workshops.
- Provide opportunities to learn about a subject during stay and after departure (via website or social media).
- Provide a range of experiences with different levels of participation. Visitors choose level of participation.
- Integrate retail, dining and entertainment as part of creative cultural district.

In addition, The Research Center of Expertise for the Experience Industry in Finland has developed a twofold approach for creating a meaningful visitor experience (Figure 1). In their framework, they argue that elements such as individuality, authenticity, story, multi-sensory perception, contrast, and interaction need to be integrated into a tourism product or activity to make it meaningful for the visitor. The more elements that are included, the more intense the perception of the experience will be, and the higher the interest from the visitor in participating in the experience. Moreover, it is also argued that the learning curve of customers will be higher and the emotional connection and potential behavioral change will be greater (LCEEI 2008).

Figure 1: Creating Meaningful Visitor Experiences[2]

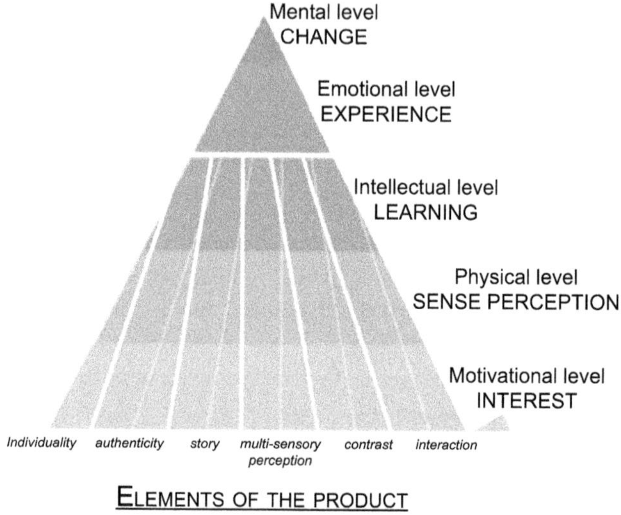

Finally, Parks Canada defines the visitor experience, "as the cumulative outcome of all aspects of each visit… including client focused services, programmes and infrastructure that support the provision of pre- and on-site trip planning information, reception and orientation, interpretation…. and the ongoing post-visit relationship" (Parks Canada 2011). These experiences provide learning opportunities, visitor satisfaction and personal connections that are personally realized, memorable and provide a shared outcome (see Table 2) (Parks Canada 2008).

Table 2: Park's Canada Visitor Experience[3]

Factors	Actions
Personally Realised	• Participatory and engaging • Based on preference and expectations • Built on quality services • Revealed over the duration of the visit • Perception of interest being met

2 LCEEI (2009). *Handbook for experience stagers*. Sanna Tarssanen (ed). Lapland Centre of Expertise for the Experience Industry. Rovaniemi, Finland.

3 Parks Canada (2008). *It's about the visitor. Visitor experience at Parks Canada*. http://www.environment.nsw.gov.au/resources/parkmanagement/CarolSheedy_May2008_TheVisitor.pdf. Accessed June 25, 2011.

Factors	Actions
Memorable	• Strong sensory interaction • Transforming • Contact with someone knowledgeable • Enjoyable • Recalled and shared with others
Shared Outcome	• Between attraction provider and visitor

The review of literature indicates that the development of creative visitor experiences requires a holistic, multi-sectoral approach that integrates experiential programming, the use of Internet communication technologies (ICT) and visitor engagement. Success requires partnering with people working in the creative industries to foster innovative approaches to develop packages that are enjoyable, transforming, provide credibility and quality service for the visitor (Wurzburger, Aageson, Pratt, & Pattakos 2009; LCEEI 2008, Parks Canada 2008).

4. Case Study: French Shore Historical Society

The following case study addresses the challenges of small rural fishing community on the island of Newfoundland on Canada's east coast, in developing creative tourism. The destination has, over time, lost their main income source, which was based on primary resources linked to the ground fishery that collapsed in the North Atlantic in the 1990s (Hull 1998). Through economic restructuring in the 21st century, the province has one of the fastest growing economies in all of Canada. This is largely as a result of offshore oil and gas development, aquaculture and commercial fishing production, mining and hydro power development and through the growth of the travel and tourism sector. In 2010, non-resident visitation grew by more than 7% to a record 518,500 visitors resulting in a 9.6% increase in expenditures totaling $410.6 million (Department of Finance 2011).

In February 2000 the French Shore Historical Society in Newfoundland and Labrador, Canada was formed as a volunteer based non-profit organization. Its initial objective was to preserve, interpret, and protect the cultural resources linked to this historical region, which was a remote summer outpost for French cod fishermen for nearly 400 years from the 16th to the early 20th century (FSHS 2011a).

Even though the French ended their fishing and use of these harbours, there are still many signs of their cultural presence, which today is the basis for a small-scale tourism industry in the remote communities of Conche, Croque, Grandois and Main Brook on the Great Northern Peninsula of Newfoundland, Canada (Figure 2).

Figure 2: Map of case study site: Island of Newfoundland, Canada[4]

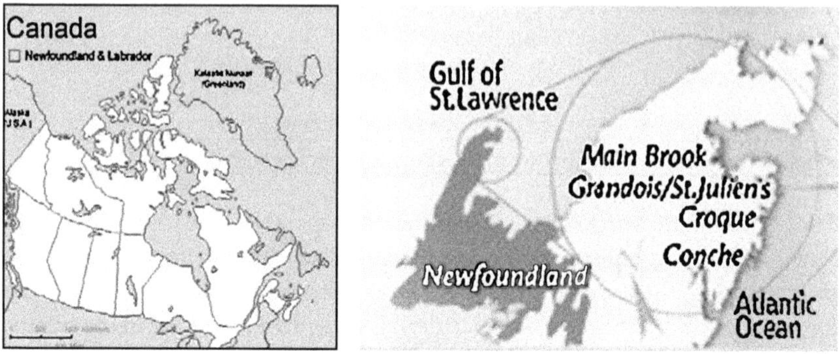

In an effort to create memorable, participatory experiences, the French Shore Historical Society has received funding from the federal and provincial government to complete a ten-year business plan for the site. As part of this plan the French Shore Historical Society is developing a number of new creative products to make connections on an emotional, physical and intellectual level with visitors linked to the traditional handicrafts practiced in the communities.

Over the past three years, local women in the community have embroidered a 222-foot long tapestry that documents the history and heritage of the region that is now on display for visitors in the French Shore Interpretation Centre (FSIC). The handicraft uses the Bayeux stitch, an Anglo-Saxon variation of an ancient technique known as laid work. The new creative tourism programming is teaching visitors the Bayeux stitch through onsite workshops. Once they learn the stitch, tourists can then work with the local women to embroider their own images of the local heritage by purchasing a kit that is for sale in the gift shop (French Shore Tapestry 2011).

The FSIC has also organized a number of different creative packages such as bread making in an original French bread oven, photography, canvas mat making, general art and painting. There are also community guided tours and archaeological excavations (FSHS 2011a).

The growth of the Center's activities are in line with the new ten year provincial tourism strategy, *Uncommon Potential*, that is responding to the growing interest of travelers in products and experiences that evolve from a unique culture and heritage off the beaten track. To achieve the desired outcome, the province

4 FSHS (2011a). At the heart of the French Shore. http://www.frenchshore.com/ Accessed April 18, 2011.

is supporting six goals for tourism development: public/private leadership; a sustainable transport network; market intelligence; tourism technology; marketing of the provincial brand and; development of the local workforce (Uncommon Potential 2011).

5. Methodology

The case study is based on a review of secondary literature examining creative- and cultural- tourism publications; an evaluation of government documents addressing the development of tourism in Newfoundland and Labrador; and web-based research of local and provincial tourism sites. In addition, a preliminary visitor survey in the communities of Conche, Croque, Grandois/St. Julien's and Main Brook was completed in June 2010 and administered by the staff of the FSIC. The main aim of the survey was to conduct an overview of the demographic and psychographic characteristics of visitors and to understand the French Shore visitor experience in a context of creative tourism.

6. Results

Out of the 2500 visitors that have travelled to the French Shore communities, 100 tourists participated in the survey providing, a 2,5% response rate. Due to the low response rate, the results must be considered with caution and have been compared to results from the Canadian Travel Activities Motivation Survey (TAMS 2003) in the discussion section to strengthen the validity of the results. A more rigorous follow up survey proposed for 2011–12 will assist in providing more valuable results for the FSIC and its marketing efforts. The main results of the survey indicate that the majority of visitors to the FSIC are tourists from Canada, with a mean age of 52. The majority of travelers have advanced degrees with an average income of $70,000CAD. They are travelling in couples, and primarily coming to the region to sightsee, hike, shop and experience the culture (Table 3).

For approximately one third of visitors, the main source of information was motivated by word of mouth (29%), followed by television (19.5%), guidebooks (15.6%), and radio (10.4%). In addition, the main reason for visiting the communities was to see the tapestry (53.1%), to the view the museum exhibits (26.5%) and to meet local staff (12.2%).

Table 3: Preliminary French Shore Interpretation Center Visitor Characteristics

Visitor	Characteristic
Non-resident Origin	Ontario (36%), Maritimes (26%), Rest of Canada (18%), USA (12%), Int'l (8%)
Motivation	Pleasure, meeting friends/relatives, business
Mean age	52
Travel	90% Couples
Education	55% University education
Income	62% CAD$70
Activities	Sightseeing, hiking, shopping, visiting historic sites

7. Discussion

The preliminary results of the survey show that visitors to the French Shore are similar to Canadian heritage enthusiasts, who are living in adult only households, and affluent with high levels of formal education. These travelers also participate in multiple heritage-related activities and are interested in the outdoors (TAMS 2003). In addition these travelers also are interested in destinations with beautiful scenery and where there are lots of things to see and do (TAMS 2003, 5). The French Shore Historical Society is focusing on tourism products that target these heritage enthusiasts. They are preserving the destination's culture and offering visitors the opportunity of being 'creatively' involved in the local traditions linked to the handicrafts of the region.

Table 4 also indicates that the province of Newfoundland and Labrador is attracting an increasing number of non-resident visitors. In 2010, the province welcomed over 500,000 visitors (Department of Tourism, Culture and Recreation, 2011). This growth in visitation is the result of an award winning national marketing campaign that focuses on the natural beauty and unique local culture. This campaign is positively impacting visitation to the French Shore communities.

Table 4: Summary of Estimates of Non-Resident Tourism Visitation Newfoundland and Labrador 2006–2010[5]

Non-Resident Tourism Visitation Newfoundland and Labrador	2006	2007	2008	2009	2010
	496,600	490,100	480,100	483,200	518,500

5 Department of Tourism, Culture and Recreation. (2011). Travel and tourism highlights 2010. Government of Newfoundland and Labrador. Department of Tourism, Culture and Recreation. St. John's, NL, Canada. http://www.economics.gov.nl.ca/E2011/TravelAndTourism.pdf Accessed June 25, 2011.

In addition, the government has also implemented a cultural tourism strategy, *Creative Newfoundland and Labrador: Blueprint for Development and Investment in Culture* (2003). The strategy acknowledges that the culture and creativity of the province is not isolated or self-contained. The report (Creative Newfoundland and Labrador 2003, 11) states:

"The creations, activities and productions of professional artists and other cultural workers, cultural industries and businesses, and cultural organizations and institutions altogether generate an extensive web of economic activity… producing (sic) approximately $289 million in output, accounting for 2% of provincial Gross Domestic Product (GDP)…"[6]

The strategy identifies that the main cultural assets are linked to the province's unique culture and traditions, and the local people and their willingness to work with tourists and to invite them to participate in their culture. The willingness of the local women on the French Shore to work with tourists is an important consideration for the development of regional tourism programming. At the FSIC it has been observed that the local women working on handicrafts are not used to working with international visitors. At first a number of the local women had to be convinced that their traditional work and knowledge was of interest to tourists. In addition, because the creative programming is a new product still being developed, tourists coming to see the tapestry are also uncertain of what to expect in terms of an experience. So far, the encounters have resulted in positive experiences on both sides, but staff training is necessary to make the women more confident in sharing their culture with visitors and in formalizing the programming onsite.

The provincial government has also identified a number of additional unmet challenges for the cultural sector in the province that are critical issues for the FSIC and the future success of its programming. They include: the weakening of some elements of traditional culture; the financial instability of cultural organizations; the impacts of globalization; insufficient capital investment; a small local market; underdeveloped cultural export; lack of accessibility; an inadequate support system for the creative process; and a shortage of cultural infrastructure (Creative Newfoundland and Labrador 2003).

At present, the French Shore Historical Society commissioned a business plan (FSHS 2011b) to address financial investment in the FSIC and its programming.

---

6 Creative Newfoundland and Labrador (2003). *Creative Newfoundland and Labrador: A Blueprint for Development and Investment in Culture.* http://www.tcr.gov.nl.ca/tcr/publications/2006/culturalplan2006.pdf. St. John's, NL.

The Executive Director and the staff are hired on a seasonal basis through government grants. There is a need for a more permanent source of funding to employ staff on a permanent basis. The community of Conche is also located approximately 20 kilometers down a dirt road making accessibility a major concern for future growth. The business plan advocates working with municipal and provincial governments to upgrade the road to increase access. Finally, the tapestry is presently on display in a room of the Interpretation Center. There is a need to build a proper exhibit space for the tapestry as well as a more formal workshop space for creative programming (FSHS 2011a).

8. Conclusions

The case study of the French Shore Historical Society identifies a number of important factors that need to be considered by tourism policymakers in developing creative tourism in rural communities. First, the main assets for creative tourism products are the unique cultural assets of the local community itself. These cultural assets serve as the basis for new programmes and activities that involve the co-production of experiences between locals and tourists. The case study raises the point that one of the major challenges for the future success of creative tourism depends upon adequate human resource training for local residents so that they do not feel inadequate in working with visitors as they share their knowledge. Due to a potential lack of familiarity and interaction with outsiders, local residents need to gain an understanding of the benefits of cross cultural experiences linked to tourism that can increase local pride and prosperity for their communities.

Second, the case study illustrates the important role of the public sector in providing technical and financial assistance for future development to cultural tourist attractions such as the FSIC. In many cases the public sector provides funding to support employment and plays a dominant role in the development process of rural regions. In the case of the FSIC, the Canadian federal and provincial governments funded a business plan for future planning and development. In general it is difficult to attract outside investment to support small and medium tourism attractions/enterprises and regional development in such insular societies. There are limited opportunities for new tourism attractions/businesses to establish themselves due to their remote, isolated locations. Local development challenges include: a narrow productive base, small populations, a large proportion of earnings spent on imports to sustain the tourism industry, the limited diversification of economic activity, far from major trade and commerce centres and lack of access. All these circumstances limit the chances for economic development (Shareef, Hoti, and McAleer 2008).

Finally, even with all these potential difficulties, the case study illustrates how local residents can enter the tourism industry and take advantage of local cultural and natural resources to generate benefits that foster sustainability. For the French Shore Historical Society, creative tourism is providing a potential economic opportunity for local residents, if it is planned and developed properly. Through greater long-term investment in human resources, programming and infrastructure at the FSIC as well as through increased accessibility, the French Shore Historical Society can continue to nurture the creative talent of the community's residents, and improve access for travelers interested in experiencing the history and culture of Newfoundland's French Shore.

References

ATLAS (2007). *Cultural tourism trends in Europe.* http://www.tram-research.com/cultural_tourism_in_europe.PDF. Accessed April 18, 2011.

Creative Newfoundland and Labrador (2003). *Creative Newfoundland and Labrador: A Blueprint for Development and Investment in Culture.* http://www.tcr.gov.nl.ca/tcr/publications/2006/culturalplan2006.pdf. St. John's, NL.

Department of Finance. (2011). The economy 2010: highlights and expectations. Government of Newfoundland and Labrador. Department of Finance. St. John's NL, Canada. http://www.economics.gov.nl.ca/E2011/HighlightsAndExpectations.pdf. Accessed June 25, 2011.

Department of Tourism, Culture and Recreation. (2011). Travel and tourism highlights 2010. Government of Newfoundland and Labrador. Department of Tourism, Culture and Recreation. St. John's, NL, Canada. http://www.economics.gov.nl.ca/E2011/TravelAndTourism.pdf Accessed June 25, 2011.

Florida, R. L. (2002). *The rise of the creative class: and how it's transforming work, leisure, community and everyday life.* Perseus Book Group. New York.

French Shore Tapestry (2011). *La tapisserie du French Shore/French Shore Tapestry.* http://www.frenchshoretapestry.com/ Accessed April 16, 2011.

FSHS (2011a). *At the heart of the French Shore.* http://www.frenchshore.com/ Accessed April 18, 2011.

FSHS (2011b). *Investing in the future: French Shore Historical Society Business Plan.* FSHS: Conche, Newfoundland.

Hull, J. (1998) *Analyzing the potential for tourism to promote sustainable economic development on the Lower North Shore of Quebec.* Ph D. thesis. McGill University. Montreal, Canada.

LCEEI (2009). *Handbook for experience stagers.* Sanna Tarssanen (ed). Lapland Centre of Expertise for the Experience Industry. Rovaniemi, Finland.

McKercher, B./du Cros, H. (2002). *Cultural tourism.* Routledge. London.

Parks Canada. (2011). *Corporate plan 2009/10-2013/14. Programme activity 4: Visitor Experience* http://www.pc.gc.ca/docs/pc/plans/plan2009-2010/sec2/page05_e.asp Accessed June 25, 2011.

Parks Canada (2008). *It's about the visitor. Visitor experience at Parks Canada.* http://www.environment.nsw.gov.au/resources/parkmanagement/CarolSheedy_May2008_TheVisitor.pdf. Accessed June 25, 2011.

Pine, J./Gilmore, J. (1999). *The Experience Economy.* Harvard Business School Press. Boston, USA.

Richards, G. (2007). Introduction: Global trends in cultural tourism. In (G. Richards editor) *Cultural tourism: global and local perspectives.* pp. 1–24. Routledge. London, England.

Richards, G./Wilson, J. (2006). *Developing Creativity in Tourist Experiences: A Solution to the Serial Reproduction of Culture?* Tourism Management 27.

Shareef, R./Hoti, S./McAleer, M. (2008) *The Economics of Small Island Tourism.* Northampton, Mass. USA: Edward Elgar Publishing.

TAMS (2003). *Tourism Activity and Motivation Study of Canadian Heritage Enthusiasts.* http://en-corporate.canada.travel/sites/Corporate/images/pdf/Research/Product-knowledge/TAMS/Canadian%20Travellers%20Culture%20and%20Entertainment/tams_report_Canadas_heritage.pdf. Accessed April 18, 2011.

Timothy, D. J./ Boyd, S. W. (2003). *Heritage Tourism.* Pearson Education Limited. Essex, England.

Uncommon Potential (2011). *Uncommon Potential: A Vision for Newfoundland and Labrador Tourism.* St. John's, NL: Government of Newfoundland and Labrador. http://www.uncommonpotential.com/ Accessed April 16, 2011.

UNWTO. (2011). *Cultural tourism trends.* http://www.unwto.org. Accessed April 19, 2011.

Wurzburger, R./Aageson, T./Pratt, S./Pattakos, A. (2009). *Creative Tourism, A Global Conversation.* Sunstone Press. Santa Fe, USA.

Brooke A. Porter

Starstruck fisherfolk: Perceptions of social importance of tourism development among artisanal fisherfolk in the Philippines

1. Introduction

Most forms of tourism involve frequent social interactions between multiple parties (e.g., commercial, participatory, observatory). Such social exchanges are often obvious and entail relatively equal participation between the host(s) and the visitor(s). However, there are other less apparent social exchanges worthy of attention. For example, while the frontline service staff of a resort may be well noticed by visitors, the behind-the-scenes staff (e.g., cleaners, maintenance) may be all but invisible to the visitors. While these indirect interactions occur frequently in the contexts of tourism, the social value of tourism has been overlooked as a driver for tourism as a development strategy (Porter & Orams, 2014).

In lesser-developed nations, remote fishing communities are often limited by isolation and further, by access to technology. Visitors to such locations are rare, even by family members; therefore, seeing a new and foreign face may be seen as a significant and noteworthy event. When considering tourism as a development strategy for such locales, it becomes important to consider the dynamics and perceived importance of potential social exchanges. Pride has been previously described as an effect of knowing a foreigner (Cole, 2006). Likewise, Towner (2013) found that in the context of surf tourism, residents from the Mentawai Islands in Indonesia were quick to describe surf tourists (new acquaintances) as their "friends."

Such coastal communities in developing countries are heavily dependent on coastal fisheries for both sustenance (Baticados, 2004; Sievanen, Crawford, Pollnac, & Lowe, 2005) and livelihoods (Allison et al., 2005; Kronen, Vunisea, Magron, & McArdle, 2010). The heavy reliance on fisheries creates considerable environmental risks for the resource and those dependent upon it (Kronen et al., 2010; Turner et al., 2007). Livelihood diversification opportunities such as those afforded by tourism development may offer some solutions to the dependency on a declining resource. To date, the majority of the literature concerning tourism as a supplemental or alternative livelihood strategy has been focused on perceptions of post-tourism integration (e.g., Cheong, 2005; Diedrich, 2007). The value of social interactions as a component of and possible driver for tourism development has been sporadically reported (see Towner, 2013; Cole 2006), yet, seldom explored. This chapter looks at a data subset from a larger study on the viability of tourism as a development strategy for remote artisanal fisherfolk. Specifically, it investigates the perceived social benefits of potential tourism development among Filipino fisherfolk. Based on findings from three cases studies (one of which was reported in detail in Orams & Porter, 2014), the perceived need for social interactions with visitors are explored as they pertain to tourism development in coastal communities in lesser-developed countries.

1.1 Development

The term "development", since its political application by U.S.A. President Truman in the late 1940s, carries decades of baggage. The associated theories and definitions of development continue to be the target of constant critique by post-development theorists (Ziai, 2007). As Esteva (1992) argues, "the word [development] indicates that one is doing well because one is advancing in the sense of a necessary, ineluctable, universal law and toward a desirable goal" (p. 10). While good will and improvement are seemingly inherent components of development, Esteva (1992), notes that for two-thirds of the world's population, the term development "is a reminder of what they are not. It is a reminder of an undesirable, undignified condition. To escape from it, they need to be enslaved to others' experiences and dreams" (p. 10). However, as described by other researchers this "enslavement" of experiences and dreams may be desirable to some (Cole, 2006; Towner, 2013). For example, it is now common to come across persons from remote locations in lesser-developed countries emulating pop-culture (author's personal observations in the Philippines).

To date, development approaches aimed at coastal fishing communities have relied heavily upon improved fisheries techniques (e.g., increasing efficiency, improving product handling/storage) or fisheries development approaches (e.g., aquaculture programmes). However, increasing attention is being given to providing alternative livelihood opportunities through development options such as tourism (e.g., Armada, White, & Christie, 2009; Bauer, 2005; Croes & Vanegas, 2008; Laws, 2009; Mensah & Amuquandoh, 2010; Mograbi & Rogerson, 2007).

For those in the developed world, life on the coast may be considered a commodity; however, for those in less-developed countries, life on the coast may be more of a necessity. The differences between such communities are observable, although the common labels assigned to measure these differences (e.g., Westernised versus non-Westernised, developed versus less developed/under-developed/

developing nations, less economically developed versus economically developed nations, industrialised versus non-industrialised, first world versus second and third world) do little to describe the actual circumstances as perceived by those being described. There are critiques associated with all these labels; however, some such as first, second and third world infer political alliances rather than size of economies or standards of living. The term "developing country" is a widely accepted label for countries that have not yet achieved their stated desired level of development and is most often used by international aid agencies such as the United Nations (UN) agencies and the World Bank. As the criterion and indexes to qualify a country as developing are fluid and vary amongst organisations, there is not an internationally recognised definition of a developing country (UN Statistics Division, 2013). Therefore, this chapter will refer to any country that is currently a recipient of development programmes or initiatives funded by international aid, with the exception of international NGOs (e.g., The Nature Conservancy, World Wildlife Fund), as a developing country or less(er)-developed nation. This definition in no way assumes that the nationals of a developing country have any desire to be developed, are accepting of 'Western' standards, or are seeking such change.

1.2 Tourism as Development

Tourism has been a long-standing suggested development strategy and it has received much attention in the literature through so called 'pro-poor tourism' efforts that suggest tourism as a poverty alleviation tool (Croes & Vanegas, 2008; Laws, 2009; UNWTO, 2004). In particular, pro-poor tourism aims to improve the livelihoods of the poor (Kenneth, 2000). However, there remain few success stories that tourism, as a development strategy, is an effective or consistent approach. Often limited to theory, the lack of widespread success may be a result of trying to bridge two often very diverse audiences (the tourists and the host community).

In the context of fisheries, there is evidence that marine tourism may provide a profitable and more sustainable use of coastal-marine resources than a resource-depleting industry such as fisheries (Garrod & Wilson, 2004; Hooker & Gerber, 2004; Orams & Forestell, 1995; Quiros, 2007). Further, nature-based marine tourism has been shown to have significant economic influence on small economies (Bauer, 2005; Beeton, 2006; Orams, 2001).

These points provide a compelling argument for tourism as a development strategy for remote fishing communities. Marine tourism continues to be an evolving economy in the Philippines with many ventures beginning as ad hoc attempts (personal observation, 2011–2014). The economic potential of tourism remains promising. A case study by Samonted-Tan et al. (2007) revealed that the net benefits of tourism in the Philippines exceed those of the fisheries. Further, their findings demonstrate that the coastal-marine environment is capable supporting more than one economic activity (Samonted-Tan et al., 2007). Similarly, White, Vogt and Arin (2000) tout the potential environmental and economic benefits (from increased catch and small-scale tourism use) from a well-managed coral reef.

Tourism, as does any strategy, has its limitations (Oracion et al., 2005). Within the Philippines, UNWTO (2006) notes significant challenges with tourism as a development strategy. These included issues such as significant delays in the processing of necessary documents due to government bureaucracy, issues with participation in training programmes due to conflicting livelihood priorities, and limited opportunities for participation in the tourism industry. Therefore, it is important to note that in some cases of tourism as a development strategy, the potential benefits may be masked by potential contradictions. Multiple works by Fabinyi (2008, 2010) and others (e.g., Goldoftas, 2006; Ong, Sotrey & Minnery, 2011) detail similar shortcomings of tourism as a development strategy in the Philippines. While he described positive linkages between MPAs and tourism in the Philippines (Fabinyi, 2008), his later research documented a case in which tourism development contributed to land disputes between developers and fisherfolk as a result of associated increases in land value (Fabinyi, 2010).

Other contradictions to tourism development include unequal wealth distribution. Pine (2007) blames corruption for the problems with wealth distribution for the locals. Similarly, Fabinyi (2010) summarises the direct effects of political interference within the Philippines, describing "many such negative effects of tourism for local fisherfolk are bound up with larger questions of class, land tenure insecurity and governance patterns in Philippine society" (p. 425). Such scenarios highlight the complexity of tourism as a development strategy. Research has shown that inequalities in the benefits from tourism growth can cause resentment towards tourism, largely due to segregation of the two sectors [tourism economy and fishing economy] that were using or competing for the same resource (Fabinyi, 2008).

Although tourism is frequently suggested as a conservation strategy (e.g., Garrod & Wilson 2004; Hooker & Gerber, 2004; Orams, 2004; Orams & Forestell, 1995), there are numerous accounts of tourism contributing to environmental degradation. For example, in the Philippines, Quiros' (2007) study, which investigated the impacts of the industry on whale shark behaviour, cited the lack of management infrastructure and enforcement activities as being detrimental to the environment. A later report by Quiros (2008) noted that in addition to environmental damages, the locals were not necessarily the ones reaping the benefits, citing leakages and government corruption as contributing factors. Similar results were found by Pine (2007) who cited politics and government red tape as a hindering factor in the Donsol whale shark tourism industry.

In addition to the potential adverse environmental and economic effects of tourism development on coastal populations, many of the target audiences are also considered a vulnerable population. In developing nations, many coastal residents are faced with poverty (Bene, 2003) and a dependence on a declining resource (Bachetta, Ernst, & Bustamante, 2009; Chowdhury, 2009). Pomeroy et al. (2006) suggest, "issues of resource access, marginalisation, market access, power imbalances, lack of information, and unsustainable resource use" as root causes of such vulnerabilities (p. 787).

1.3 The Philippines

The Republic of the Philippines is an archipelago in the Western Pacific Ocean of over 7,100 islands. To the North it is bordered by the South China Sea and to the South, the Sulu Sea. The archipelago falls within the coral triangle and is considered to be the "centre of the centre of marine shore fish biodiversity" (Carpenter & Springer, 2005, p. 467). Though remarkable, mismanagement and misuse of the marine environment is rampant throughout the Philippines (Baticados, 2004; Baticados & Agbayani, 2000; Dalabajan, 2009; Fabinyi, 2010; Johnstone, Labrado, Eisma-Osorio, Christie & White, 2010; Oracion et al., 2005; Samonted-Tan et al., 2007). The Philippine Islands are divided into three main regions from north to south (Luzon, Visayas and Mindanao) each of which is further divided into administrative divisions, provinces, municipalities and barangays. Barangays serve as the "fundamental territorial unit and political building block of Filipino society" and vary in size from 100 to around 2000 persons (Oracion et al., 2005, p. 396). The population of the Philippines is estimated at over 105 million people (Central Intelligence Agency, 2013) with over two thirds of these residing in coastal communities (The World Bank, 2005). However, it is expected that the population is likely much higher due to unrecorded births in remote areas.

Weather plays a significant role in the fisheries of the Philippines. The country is subject to two annual typhoon or monsoon seasons, *habagat* [the southwest monsoon] and *amihan* [the northeast monsoon]. *Amihan* occurs between the cooler months of November to May, while the wetter *habagat* season fills the remainder of the year. It is common for over twenty typhoons to hit the country each year. While effects of larger typhoons are obvious, weather systems during the *amihan* season, cause many of the smaller vessels, common to the artisanal fisheries, to remain grounded for days and sometimes weeks (Fabinyi, 2010; Goldoftas, 2006).

2. Rethinking Development

The rapidly growing population of the Philippines coupled with the unsustainable use of and dependency on the marine resources (Cruz-Trinidad, Geronimo, & Alino, 2009) leaves many Filipinos vulnerable (Johnstone et al., 2010). This vulnerability creates a need for change and development within the Philippines. Fisheries development projects have come under scrutiny for increasing pressure on the fisheries (which are often in decline) while failing to produce the social goals (Hauck, 2007; Hollup, 2000; Muallil et al., 2011). In line with the original credos of post-development theorists (e.g., Esteva, 1992; Sachs, 1992), the cases discussed in this chapter realise the need for change to the current approaches to development. This is achieved through a lesser-explored angle of pre-development social desires of a specific audience (remote artisanal fisherfolk). This discussion is grounded in perception-based data regarding perceived social benefits of tourism development of remote artisanal fisherfolk in the Philippines.

The case study has been suggested as a method for furthering tourism research (Beeton, 2005). Data from three cases, what Stake (1995) refers to as a "collective case study" were used to explore the potential (or lack thereof) for tourism as a supplemental or alternative livelihood for remote artisanal fishing communities in a developing nation, the Philippines. While it is a long-standing belief that findings from case studies are limited in transferability, Stake (1978) argues that often the situation "is one which there is a need for generalisation about that particular case or generalisation to a similar case rather than generalisation to a population of cases" (p. 8). This is supported by Flyvbjerg (2006) who argues that cases are indeed transferable describing the consequence of a critical case, or one that completely dispels existing theory.

This chapter explores the role of social benefits of potential tourism development using perceptions of fisherfolk from three Filipino communities to suggest methods of social inclusion for future development efforts. The cases described all have similar characteristics associated with target audiences for development projects. The majority of the community members were living in poverty according to government standards. The most common dwellings consist of simple structures made from a mix of wood, woven mats, concrete blocks, and/or corrugated tin. Dwellings are covered with a roof of woven palm fronds or corrugated tin. Dirt-floors are common. Communication is largely done via Short Message Service (SMS) on cell phones, which is affordable (less than two US cents per message); it is common for a household to share a single cell phone. Traditionally, the men go out to sea to fish, while the women tend to family needs and participate in side-businesses, known locally as sideline activities, such as gleaning (also referred to as beach seining) or mat weaving.

2.1 The Communities

Each case or unit of analysis was bounded by a *barangay*. The first inquiry took place in the municipality of Bolinao in Barangay Victory and involved twenty-one participants. The second inquiry took place in the Coron municipality in Barangay Decabobo and involved eight participants. The third inquiry took place in the Busuanga municipality, which is adjacent to Coron, in Barangay Dimipac and involved fourteen participants.

The perceived effects of potential tourism development on the community were explored through interview questions that sought to define the costs and benefits of tourism development to the community (see Table 1). Particularly, questions attempted to define perceived social benefits of tourism as well as any potential ethical risks to the community (e.g., changes in culture, prostitution). Participants did not report any potential negative effects on the community associated with tourism development. Many felt that tourism development would have "no effect" on the community. Some (n=6) participants admitted to not understanding or knowing how or if tourism could affect the community. Social benefits overshadowed potential economic benefits. Positive social effects included a fun atmosphere associated with visitors, the chance to share their place and the ability to see new faces. A recurring theme was the joy or excitement of seeing a new face. As one woman stated:

> I remember when we were kids; we were fascinated to see foreigners. We loved to see foreigners. I cannot explain why we were starstruck. I guess we were amazed that someone from another part of the world would come here.

Table 1: Perceived costs and benefits of tourism on the community

Location	**No effect**	**Positive effect**	
		Economic	**Social**
Victory	9	3	2
Decabobo	7	0	0
Dimipac	5	0	9
Total	**21**	**3**	**11**

Feelings towards visitors were also explored during the interview sessions (see Table 2). With the exception of one comment representing two secondary participants, all other interviewees reported feelings toward potential visitors as either indifferent or positive. Many reported a general happiness at the idea of welcoming a visitor. Some participants provided justification to the feelings of happiness. As one explained, "we are happy because new faces come here." The abilities to see new faces, meet new people (as well as potentially find a spouse) and hear news were themed as a 'change of pace.' This break from the day-to-day routine was the most common response. One participant in his early twenties speculated on a chance at wealth and love saying, "it's ok for them [the tourists] to come. We can marry a tourist; jackpot – then you get the dollars!" This statement was said partially as a joke, but it is also recorded as a true representation of the social possibilities associated with visitors. Pride was a sentiment mentioned solely by Victory fisherfolk. Another nuance in the data that was limited to the Victory research site was mention of the researcher's personal appearance. Four men specifically stated that they would welcome "beautiful visitors [in reference to the researcher's appearance]".

Two elderly women in Dimipac conveyed hesitation about the effects of visitors on culture with one stating (and the other agreeing):

> The tourism affects our community and culture. Most of the tourists go to the dance place in the community. They wear shorts and the tradition is more formal. They should wear pants or skirt, but the young generation has changed. We see lots of foreigners wearing bikinis. Before it's not good to look at them, but now it is commonplace.

While this response is indicative of an acceptance of change, it identifies a discomfort associated with the presence of outsiders.

Participant responses grouped as indifferent included those that said, "we feel nothing." Additionally, another participant noted the racial differences of visitors, explaining, "those are the white people. Those are the people who have white colours and we do not. It's like that. They are white. It's ok if they come to visit." Another participant stated that, although there was tourism nearby, they did not feel its effects (positive or negative). Yet another interviewee, when asked to expand on his feelings towards the tourists, responded saying "they [the visitors] are humans; they want to enjoy." In such cases when there was no indication of positive or negative feelings, these responses were classified as indifferent. Indifference was not reported from Victory participants.

Other respondents associated improvement of livelihoods through infrastructure or diversification with welcoming visitors. Specific mention was given to new livelihood opportunities such as being able to sell products to visitors or hire out boats. One participant summarised multiple positive effects of visitors, stating:

> They [visitors] are beautiful. I am happy and proud because we expect the unexpected. We think foreigners will help the livelihoods by giving some projects and help uplift the economy and financial status. There are only a few foreigners coming. We want to see new faces, new looks, a new colour, and make new friends. In these barrios [neighbourhoods] we just look at each other. Every day, every time, it's the same face.

Table 2: Feelings towards visitors

Negative effects	Number of responses			Indifferent effects	Number of responses			Positive effects	Number of responses		
	Victory	Decabobo	Dimipac		Victory	Decabobo	Dimipac		Victory	Decabobo	Dimipac
Changes culture	0	0	2	Indifference	0	3	6	General happiness	10	5	3
								Change of pace	14	3	9
								Feelings of pride	5	0	0
								Improve livelihoods	2	0	12

3. Perception-based Priorities

Consistent throughout the research sites, some participants specifically referred to the joy of "seeing a new face" and the ability to "hear news". These types of responses were coded under "change of pace." Little attention has been given to the social desires of fisherfolk. Thus, the findings offer significant insights to previously unexplored areas for indicators or drivers of development. Based on the emphasis given to seeing new faces, making friends or hearing news, in the context of remote fishing communities, it was interpreted that a tourist or visitor is associated with a break in the monotony of day-to-day life where external stimulation (e.g., cell phone, internet, television, radio) may be limited or non-existent. Such desires for social interactions have been seemingly missed by previous literature which commonly refer to fisherfolk as a somewhat reclusive demographic (e.g., Muallil et al., 2009; Pollnac et al., 2001; Smith, 1981).

The stated willingness of fisherfolk to interact with visitors bodes well for tourism as a development strategy. However, what remains unknown is how such actions would take place. Personal observations from the field revealed that many of the fisherfolk participants were somewhat shy, and further, only a few spoke enough English to facilitate a rudimentary communication. Thus, it is expected that the lack of hospitality skills and language skills observed within this demographic would influence tourism outcomes. However, Porter & Orams (2014) suggest, “voyeuristic requests… may be satisfied through an indirect exposure to tourism and may not require a direct engagement in the industry” (p. 18). Tourism, while far from a simple strategy, has the potential to provide benefits to the community while lessening the dependency on the marine resources.

4. Looking Ahead

When considering tourism as a supplemental or alternative livelihood, the success is dependent not only on the feasibility of integration (e.g., the knowledge required and the ability to provide reasonable services demanded by tourism), but also on the given communities desire to engage in the industry. In the case of remote artisanal fishing communities in developing nations, the lack of exposure to tourism and to outsiders creates significant hurdles for a livelihood shift towards tourism. While further research is necessary to determine the transferability of the findings, shifting the focus of current development strategy from monetary and/or environmental benefits to social benefits would delay the sometimes urgent goals of livelihood improvement and lessening the dependence on the resource. However, acknowledging this social desire within some remote fishing communities may serve to improve the long-term outcomes of development programmes which to date have shown questionable results.

References

Allison, E. H., Adger, W. N., Badjeck, M., Brown, K., Conway, D., Dulvy, N. K., … Reynolds, J. D. (2005). *Effects of climate change on the sustainability of capture and enhancement fisheries important to the poor: analysis of the vulnerability and adaptability of fisherfolk living in poverty* (R4778J). Retrieved from Fisheries Management Science Programme Department website: http://www.fmsp.org.uk/Documents/r4778j/r4778j_1.pdf.

Armada, N., White, A. T., & Christie, P. (2009). Managing fisheries resources in Danajon Bank, Bohol, Philippines: An ecosystem-based approach. *Coastal Management, 37*(3–4), 308–330. doi:10.1080/08920750902851609

Baticados, D. B. (2004). Fishing cooperatives' participation in managing nearshore resources: The case in Capiz, central Philippines. *Fisheries Research, 67*(1), 81–91. doi:10.1016/j.fishres.2003.07.005

Baticados, D. B., & Agbayani, R. F. (2000). Co-management in marine fisheries in Malalison Island, central Philippines. *International Journal of Sustainable Development and World Ecology, 7*(4), 343–356. doi:10.1080/13504500009470053

Bauer, J. J. (2005). *Conservation, poverty alleviation and community development through tourism in developing countries: The international program of the CRC for sustainable tourism (1999–2002)* (1). Gold Coast, Australia: CRC Sustainable Tourism.

Beeton, S. (2005). The case study in tourism research: A multi-method case study approach. In B. W. Ritchie, P. Burns, & C. Palmer (Eds.), *Tourism research methods: Integrating theory with practice* (pp. 37–48). Cambridge, MA: CABI Publishing.

Beeton, S. (2006). *Community development through tourism.* Collingwood, Australia: CSIRO Publishing.

Carpenter, K. E., & Springer, V. G. (2005). The center of the center of marine shore fish biodiversity: The Philippines Islands. *Environmental Biology of Fishes, 72*(4), 467–480.

Cheong, S. (2005). Korean fishing communities in transition: Limitations of community-based resource management. *Environment & Planning, 37*(7), 1277–1290. doi:10.1068/a37139

Cole, S. (2006). Information and empowerment: The keys to achieving sustainable tourism. Journal of Sustainable Tourism, 14(6), 629–644.

Croes, R., & Vanegas, M. (2008). Cointegration and causality between tourism and poverty reduction. *Journal of Travel Research, 47*(1), 94–103. doi:10.1177/0047287507312429

Cruz-Trinidad, A., Geronimo, R. C., & Alino, P. M. (2009). Development trajectories and impacts on coral reef use in Ligayen Gulf, Philippines. *Ocean & Coastal Management, 52*(3), 173–180. doi:10.1016/j.ocecoaman.2008.12.002

Dalabajan, D. (2009). Of crimes and no punishments: Fisheries law offenses and the criminal justice system in the Calamianes group of islands. In R. D. Anda, & D. A. Dalabajan (Eds.), *Against the Tide: Enforcement and Governance in the Sulu Sulawesi Seas* (pp. 50–64). Puerto Princesa City, Philippines: Environmental Legal Assistance Center.

Diedrich, A. (2007). The impacts of tourism on coral reef conservation awareness and support in coastal communities in Belize. *Coral Reefs, 26*(4), 985–996. doi:10.1007/s00338-007-0224-z

Esteva, G. (1992) Development. In W. Sachs (Ed.) *The Development Dictionary. A Guide to Knowledge as Power,* (pp. 6–25). London, UK: Zed Books.

Fabinyi, M. (2008). Dive tourism, fishing and marine protected areas in the Calamianes Islands, Philippines. *Marine Policy, 32*(6), 898–904. doi:10.1016/j.marpol.2008.01.004

Fabinyi, M. (2010). The intensification of fishing and the rise of tourism: Competing Coastal livelihoods in the Calamianes Islands, Philippines. *Human Ecology, 38*(3), 415–427. doi:10.1007/s10745-010-9329-z

Flyvbjerg, B. (2006). Five misunderstandings about case-study research. *Qualitative Inquiry, 12*(2), 219–245. doi:10.1177/1077800405284363

Garrod, B., & Wilson, J. C. (2004). Nature on the edge? Marine ecotourism in peripheral coastal areas. *Journal of Sustainable Tourism, 12*(2), 95–120. doi:10.1080/09669580408667227

Goldoftas, B. (2006). *The green tiger: The cost of ecological decline in the Philippines.* New York, NY: Oxford Press.

Hauck, M. (2007). Rethinking small-scale fisheries compliance. *Marine Policy, 32*(4), 635–642. doi:10.1016/j.marpol.2007.11.004

Hollup, O. (2000). Structural and sociocultural constraints for user-group participation in fisheries management in Mauritius. *Marine Policy, 24*(5), 407–421. doi:10.1016/S0308-597X(00)00016-6

Hooker, S. K., & Gerber, L. R. (2004). Marine reserves as a tool for ecosystem-based management: The potential importance of megafauna. *BioScience, 54*(1), 27–39. doi:10.1641/0006-3568(2004)054[0027:MRAATF]2.0.CO;2

Johnstone, P., Labrado, G., Eisma-Osorio, R.-L., Christie, P., & White, A. (2010). *Marine protected area networking training handbook.* Cebu City, Philippines: Coastal Conservation and Education Foundation, Inc.

Kenneth, F. H. (2000). Recognising deductive processes in qualitative research. *Qualitative Market Research: An International Journal, 3*(2), 82–90. doi:10.1108/13522750010322089

Kronen, M., Vunisea, A., Magron, F., & McArdle, B. (2010). Socio-economic drivers and indicators for artisanal coastal fisheries in Pacific island countries and territories and their use for fisheries management strategies. *Marine Policy, 34*(6), 1135–1143. doi:10.1016/j.marpol.2010.03.013

Laws, E. (2009). ProPoor tourism – who benefits? Perspectives on tourism and poverty reduction. *Tourism Management, 30*(1), 143–145. doi:10.1016/j.tourman.2008.05.002

Mensah, E. A., & Amuquandoh, F. E. (2010). Poverty reduction through tourism: Residents' perspectives. *Journal of Travel and Tourism Research* (Spring 2010), 77–96.

Mograbi, J., & Rogerson, C. M. (2007). Maximising the local pro-poor impacts of dive tourism: Sodwana Bay, South Africa. *Urban Forum, 18*(2), 85–104. doi:10.1007/s12132-007-9002-9

Muallil, R. N., Geronimo, R. C., Cleland, D., Cabral, R. B., Doctor, M. V., Cruz-Trinidad, A., & Alinio, P. M. (2011). Willingness to exit the artisanal fishery as a response to scenarios of declining catch or increasing monetary incentives. *Fisheries Research 111*(1), 74–81. doi:10.1016/j.fishres.2011.06.013

Ong, L. T. J., Storey, D., & Minnery, J. (2011). Beyond the beach: Balancing environmental and socio-cultural sustainability in Boracay, the Philippines. *Tourism Geographies 13*(14), 549–569. doi:10.1080/1416688.2011.590517

Oracion, E. G., Miller, M. L., & Christie, P. (2005). Marine protected areas for whom? Fisheries, tourism, and solidarity in a Philippine community. *Ocean & Coastal Management, 48*(3), 393–410. doi:10.1016/j.ocecoaman.2005.04.013

Orams, M. B. (2001). From whale hunting to whale watching in Tonga: A sustainable future? *Journal of Sustainable Tourism, 9*(2), 128–146. doi:10.1080/09669580108667394

Orams, M. B. (2004). The use of the sea for recreation and tourism. In H. D. Smith (Ed.), *The oceans: Key issues in marine affairs* (pp. 161–173). Netherlands: Springer. doi:10.1007/978-1-4020-2780-2_9

Orams, M. B., & Forestell, P. H. (1995). From whale harvesting to whale watching: Tangalooma 30 years on. In O. Bellwood, H. Choat, & N. Saxena (Eds.), *Recent Advances in Marine Science and Techonology '94* (pp. 667–673). Townsville, Australia: James Cook University of North Queensland.

Pine, R. (2007). *Donsol whale shark tourism and coastal resource management – A case study on the Philippines.* Retrieved from World Wildlife Fund Philippines website: http://www.wwf.org.ph/downloads/DonsolCaseStudy.pdf

Pollnac, R., & Pomeroy, R., Harkes, I. (2001). Fishery policy and job satisfaction in three Southeast Asian fisheries. *Ocean & Coastal Management, 44*(7), 532–544. doi:10.1016/S0964-5691(01)00064-3

Porter, B. A., & Orams, M. B. (2014). Exploring tourism as a potential development strategy for an artisanal fishing community in the Philippines: The case of Barangay Victory in Bolinao. *Tourism in Marine Environments, 10*(1–2), 49–70.

Quiros, A. (2007). Tourist compliance to a Code of Conduct and the resulting effects on whale shark (*Rhincodon typus*) behavior in Donsol, Philippines. *Fisheries Research, 84*(1), 102–108. doi:10.1016/j.fishres.2006.11.017

Quiros, A. (2008). Overview of whale sharks and whale shark tourism and whale shark tourism in the Philippines. *Proceedings of the 2nd International Whale Shark Conference, Holbox, Quintana Roo, Mexico.* Retrieved from http://www.domino.conanp.gob.mx/doc conf/Angela.pdf.

Sachs, W. (1992). Introduction. In W. Sachs (Ed.) *The Development Dictionary. A Guide to Knowledge as Power,* (pp. 1–5). London, UK: Zed Books Ltd.

Samonted-Tan, G. P. B., White, A. T., Tercero, M. A., Diviva, J., Tabara, E., & Caballes, C. (2007). Economic valuation of coastal and marine resource: Bohol Marine Triangle, Philippines. *Coastal Management, 35*(2), 319–338. doi:10.1080/08920750601169634

Sievanen, L., Crawford, B., Pollnac, R., & Lowe, C. (2005). Weeding through assumptions of livelihood approaches in ICM: Seaweed farming in the Philippines and Indonesia. *Ocean and Coastal Management 48*(3), 297–313. doi:10.1016/j.ocecoaman.2005.04.015

Smith, I. R. (1981). Improved fishing incomes when resources are overfished. *Marine Policy, 5*(1), 17–22. doi:10.1016/0308-597X(81)90070-1

Stake, R. E. (1995). *The art of case study research.* Thousand Oaks, CA: SAGE.

Towner, N. (2014). Surf tourism and sustainable community development in the Mentawai Islands, Indonesia: A multiple stakeholder perspective (PhD thesis). Retrieved from http://hdl.handle.net/10292/7149.

Turner, R. A., Cakacaka, A., Graham, N. A. J., Polunin, N. V. C., Pratchet, M. S., Stead, S. M., & Wilson, S. K. (2007). Declining reliance on marine resources in remote South Pacific societies: Ecological versus socio-economic drivers. *Coral Reefs, 26*(4), 997–1008. doi:10.1007/s00338-007-0238-6

UN Statistics Division. (2013). *Composition of macro geographical (contiental) regions, geographical sub-regions, and selected economic and other groundings.* Retrieved from http://unstats.un.org/unsd/methods/m49/m49regin.htm#ftnc.

UNWTO. (2004). *Tourism and poverty alleviation: Recommendations for action* (1). Madrid, Spain: United Nations World Tourism Organization. Retrieved from http://ezproxy.aut.ac.nz/login?url=http://YU7RZ9HN8Y.search.serialssolutions.com/?V=1.0&L=YU7RZ9HN8Y&S=JCs&C=TC_003011953&T=marc.

White, A., Vogt, H., & Arin, T. (2000). Philippines Coral Reefs Under Threat: The economic losses caused by reef destruction. *Marine Pollution Bulletin, 40*(7), 598–605. doi:10.1016/S0025-326X(00)00022-9

Ziai, A. (2007). Concluding the exploration. Post-development reconsidered. In A. Zaia (Ed.) *Exploring Post-development. Theory, practice, problems and perspectives,* (pp. 226–234). New York, NY: Routledge.

Tourism and the Environment

Marinette Kruger, Peet van der Merwe and Melville Saayman

Residents' Perceptions Regarding an Urban National Park in South Africa

1. Introduction

One of South Africa's most important role-players in wildlife-based tourism at present is South African National Parks (SANParks). SANParks is currently managing 21 national parks, each representing a unique biological diversity of biomes, plants and animal species and it is a major ecotourism attraction for local and foreign tourists, with 4.5 million tourists going through SANParks' gates during the 2010/11 season (Stevens, personal communication). South African National Parks showed a steady increase of visitors from 3.6 million (2003/04 season) to 4.5 million (2010/11 season). When looking at these figures, it is clear that national parks in South Africa attract a large number of tourists each year (Stevens, personal communication). With this in mind, national parks have three distinct purposes. Firstly, the conservation of a representative sample of the biodiversity of the country; secondly, to provide a recreation outlet where the wonders of the parks can be experienced and enjoyed, and thirdly, to maintain a relationship of community uplifting and capacity building among people living in and around the parks (Saayman & Saayman, 2006). National parks therefore have to create an environment from which the communities can benefit while at the same time conserving the natural environment (Saayman, Van der Merwe, Saayman & Mouton, 2009). According to The Office of Environment and Heritage in New South Wales, there are different types of national parks, for example traditional, marine and urban parks. This paper will, however, focus on an urban national park, namely the Table Mountain National Park (hereafter referred to as TMNP) (Office of Environment and Heritage, 2011).

Urban refers to an area that is densely populated by humans (Bryant, 2006) and, according to Chiesura (2004) and Salazar and Menendez (2007), an urban park, also known as a municipal park or a public park, public open space or municipal gardens, is a park in cities and other incorporated places to offer recreation and green space to residents of, and visitors to, the municipality, such as TMNP. Gobster (2001) and Thompson (2002) add that urban parks are also viewed as spaces where a spectrum of recreational and leisure activities can be pursued, from active endeavours to passive activities such as walking, picnicking and relaxing, and they are seen as places that provide people with unique experiences as a result of the natural and cultural features present and the social communities that gravitate to them. However, this definition and description does not encompass the attributes of an urban national park such as TMNP, which has an underlying conservation principle. Walker and Walker (2011) and Cook, Yale and Marqua (2010) define a national park as a large natural place that has a wide variety of attributes, sometimes including historical assets, while Page and Connell (2009) indicate that a national park is a designated area where the protection of natural resources is the priority and where recreation and tourism uses are accommodated through sensitive management. Based on the aforementioned, an urban national park can be defined as a large natural area situated in or near an urban area that has a variety of recreational and tourism activities that are managed in a sensitive manner.

The TMNP, is situated at the south-western tip of Africa, and stretches from Signal Hill in Cape Town all the way south to South Point. The Park includes the following areas: Table Mountain, Tokai forest, Boulders Beach and Cape Point (Cape of Good Hope) (sections framed in Figure 1). This Park was officially established in 1998 and since then has expanded to 25,000 hectares of land. Its jurisdiction stretches over 1,000km^2 of seas and coastlines around the peninsula (SANParks, 2010). The TMNP is a scenic park offering both mountain and ocean views. The Park consists of a narrow piece of land that encompasses beautiful valleys, bays and beaches that are surrounded by the Atlantic Ocean. Two world-renowned landmarks are found here, namely the world-famous Table Mountain and the legendary Cape of Good Hope. The Park is also a World Heritage Site and is famous for its diverse and unique fauna and flora, its rugged cliffs, steep slopes and sandy flats, its natural beauty, rich history, and its cultural as well as recreational assets. Activities at the Park include hiking, overnight trails, fishing, extractive diving, dog walking, scuba diving, mountain biking, surfing, wind surfing, rock climbing, picnic and braai areas, hang gliding, paragliding, and horse riding. TMNP was furthermore voted by the public as the Best Attraction in Cape Town in 2009 (SANParks, 2009).

Figure 1: Table Mountain National Park (Source: Cape Town Maps, 2011)

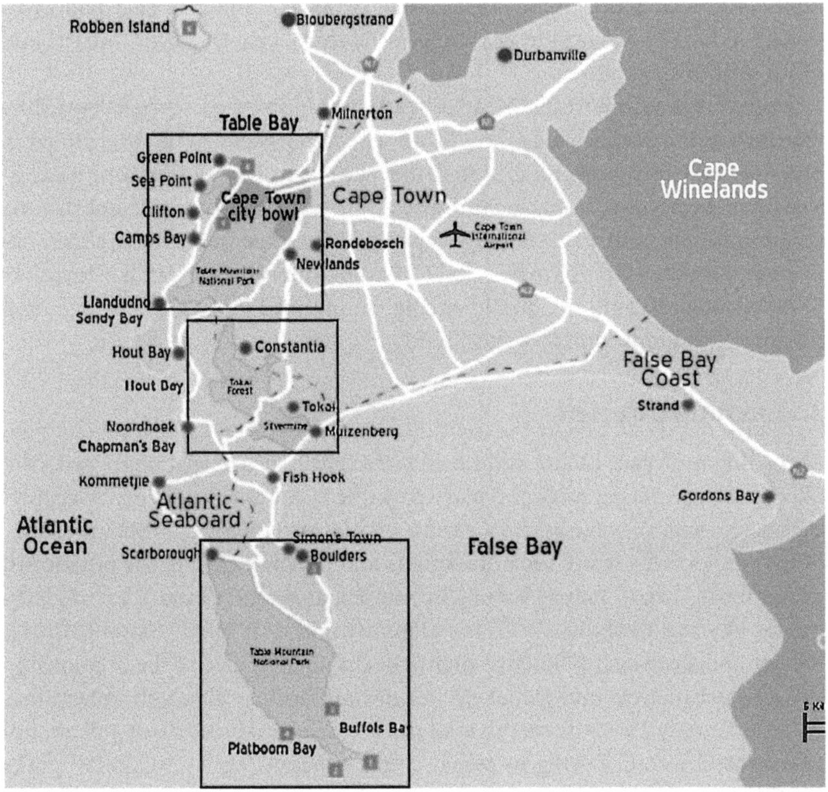

During the 2009 season, TMNP attracted more than 2.2 million visitors – almost half of the total number of visitors to all the 21 national parks in South Africa combined (Stevens, personal communication). Also taking into consideration the fact that TMNP is an urban national park, located within the city and surrounding regions of Cape Town, it is inevitable that the Park and the tourists will have some sort of socio-economic impact on the various communities and therefore the question that arises is what the perceptions of the residents regarding the social impact of TMNP are. This is because the host community, according to Akama and Kieti (2007), is considered to be one of the most important role-players in determining the success of tourism. Ratz (2000) argues that an understanding of residents' perceptions can help local decision-making as well as to improve negative perceptions and aspects of concern raised by residents. Oberholzer, Saayman, Saayman and Slabbert (2009) emphasise the importance of the latter and indicate that bad community relations can lead to poaching, vandalism, crime against tourists and other unlawful activities.

Therefore, it is important to understand communities' perceptions regarding the impact of a park such as the TMNP. The latter needs to be determined in order to make recommendations such that the positive impacts may be maximised, and the negative impacts minimised. It is furthermore important that the communities develop pride and live a quality life, as this will influence their behaviour towards tourists (SANParks, 2010). The aim of this research is therefore to determine residents' (of neighbouring communities) perceptions regarding the socio-economic impact of the TMNP.

2. Literature Review

According to Bryant (2006) and Lo and Jim (2010), urban (national) parks are effective conservation tools and play a key role in urban ecosystems by supplying a wide range of services and amenities. Chiesura (2004) adds that urban (national) parks fulfil many social functions and psychological needs of citizens, which makes urban nature a valuable municipal resource, and a key ingredient for city sustainability. With the aforementioned in mind, Grimm, Grove, Pickett and Redman (2000) also highlight the significance of the relationship between urban areas and global environmental health – although urban areas account for only 2% of the earth's land surface, they produce 78% of greenhouse gases, thereby contributing to global climate change. Urban (national) parks are therefore crucial for the economic health of cities and residents' quality of life for several reasons. Urban (national) parks make an important contribution to public health by removing or ameliorating the presence of pollutants from the air (Salazar & Menendez, 2007; Handy *et al.*, 2002; Frank *et al.*, 2003; Humpel *et al.*, 2004; Hoehner *et al.*, 2005).

Previous research also indicates associations between nature and stress responses. More frequent visits to urban (national) parks were significantly related to fewer self-reported experiences of stress (Grahn & Stigsdotter, 2003; Gidlöf-Gunnarsson & Öhrström, 2007). Kuo and Sullivan (2001) even showed that residents living in "greener" surroundings report lower levels of fear, fewer incivilities, and less aggressive and violent behaviour. Urban (national) parks furthermore contribute notably to a healthy outdoor living environment since they provide residents with outdoor recreational opportunities and contacts with nature to nurture a harmonious people-environment relationship (Kaplan

& Kaplan, 1989; Miller, 1997; Jim, 2004). Kaplan (2001) explains that for many urban residents, everyday life is filled with stressful situations. Demands from work and family and harmful/adverse environmental conditions raise the need of restoration periods, and increasing empirical evidence indicates that nature and urban parks provide restorative experiences that directly affect people's psychological well-being and health in positive ways. In addition, aesthetic, historical and recreational values of urban (national) parks increase the attractiveness of the city and promote it as tourist destination, thereby generating employment and revenue (Chiesura, 2004). The function of urban nature can therefore provide economic benefits for both municipalities and residents. Mouton (2010) and Oberholzer (2010), who conducted research on the Wilderness National Park (which is also an urban park) and Tsitsikamma National Park in South Africa, respectively, regarding the socio-economic impact of the parks, both found that the parks had a bigger social impact than economic impact on the local community. Both studies showed that a national park plays an important role in uplifting local communities' well-being.

With the above-mentioned in mind, Hetzer (1965), Blamey (2001) and Van der Merwe and Saayman (2004) identified four pillars of ecotourism (Figure 2), namely the conservation and promotion of the natural and cultural environment; the sustainable management of the environment; participation of the local community; and tourists' satisfaction.

Figure 2: Pillars of ecotourism (Adapted from Geldenhuys, 2009:3)

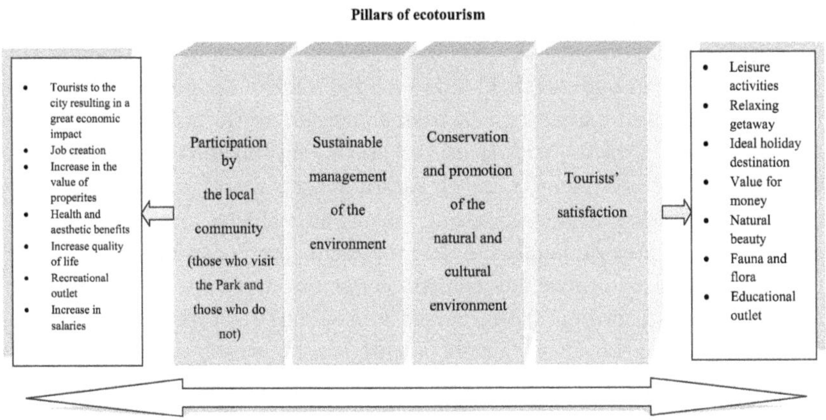

As shown in Figure 2, from a tourism perspective, urban national parks such as the TMNP in South Africa provide tourists with a nature and leisure experience and are major drawing cards for domestic travellers and especially international visitors who travel to the country for a nature experience (Saayman & Saayman, 2009). Urban national parks are therefore powerful magnets to tourists, major export earners and constitute an important part of the tourism industry (Uysal, McDonald & Martin, 1994; Cook, Yale & Marqua, 2010). From a community perspective, urban national parks are furthermore seen as a tourism product that has the capacity to contribute to job creation in the surrounding communities because conservation is labour intensive and has the ability to generate significant employment and help with poverty alleviation. This leads to increases in salaries, creates significant recreational and aesthetic value and makes a contribution to the quality of life of residents and communities neighbouring national parks (Standish, Boting, Van Zyl, Leiman & Turpie, 2004). Mouton (2010) confirms the just mentioned and states that the Wilderness National Park added to the appearance of the area, property value, social and moral values, and pride among the community has increased. In her study regarding the Tsitsikamma National Park, Oberholzer (2010) adds that residents who visited the Park more frequently do have a more positive attitude towards the Park. Tayor, Gough, Warren and McClintock (1999) contend that the social impact of urban national parks includes the shift from the production to service industry, a change in the nature and composition of employment in the area, a more viable medical and transport service, and a general change in the lifestyles of the people surrounding the area. Urban (national) parks therefore have a variety of benefits; however, the four pillars need to work in cohesion to effectively capitalise on these benefits.

Urban (national) park managers are therefore continually planning to improve their parks (Taplin, Scheld & Low, 2002). One of the greatest challenges to urban parks and managers is to balance tension between providing for the diverse uses and values of park space and preserving and enhancing the unique qualities of the place (Gobster, 2001). Park managers therefore tend to concentrate on the physical resource, valuing it as wildlife habitat, conservation land, or as scenic or historic landscape. Improvements are consequently driven by perceived needs to improve biodiversity, stem erosion, protect scenic values, or restore historic structures (Taplin, Scheld & Low, 2002). Gobster (2001) indicates that eco-system approaches offer the potential for large urban parks to be more aesthetic symbols of nature – such parks can contribute to local and regional biodiversity, provide critical nodes and linkages to maximise habitat values, and bring nature education experiences in close proximity for millions of people.

However, if such approaches are to succeed, urban (national) park managers must also work to incorporate the diverse cultural values that people hold towards urban nature (Gobster, 2001). This is because the host community plays an important role in the tourism industry and is one of the most important role-players in determining the success of tourism (Fredline & Faulkner, 2001; Fredline Jago & Deery, 2003). More specifically, there needs to be a distinction of the views of those residents who visit urban (national) parks and those who do not. West et al. (2006) explain that ecotourism activities in protected areas, such as urban (national) parks, create the expectation of economic benefit to communities surrounding the area and that the socio-economic benefit often depends on the extent of local community involvement in the activities of the park. Ratz (2000) adds that tourism developments vary from significant improvement and increase of host communities to significant deterioration and decrease of host communities and that the host communities perceive tourism's impacts in a different way than the tourists do. Tourists' perceptions mainly focus on changes that affect their holiday experience, whereas the host community's perceptions covers all areas of their own quality of life. With this in mind, Adams and Hutton (2007), and West et al. (2006) point out that in addition to the issue of displacement, the restriction of communities to inhabit and live from the land, as well as direct costs such as damages to local crops and human deaths caused by wild animals, and the change in the way local communities view themselves and their environment, are seen as some of the most noteworthy social effects of protected areas. Grahn (1985) and Bixler and Floyd (1997) also indicate that besides positive effects, parks may also have a negative effect on people's perceptions, including residents' feelings of insecurity associated with vandalism, and fear of crime in deserted places.

Owing primarily to widespread park-people conflicts and taking a cue from the world-wide trend in participatory management, Mehta and Heinen (2001) suggest community-based conservation. According to Walker and Walker (2011), community-based tourism focuses on community involvement, which generates benefits to local communities and therefore ensures that a majority of the benefits go to locals, and not outsiders. Broad-based community involvement in the planning and implementation of restoration efforts is as critical as interdisciplinary professional involvement; it empowers stakeholders and helps ensure that the landscapes they desire will be maintained over the long term (Gobster, 2001). Ratz (2000) states that understanding residents' perceptions may help local decision-making, and may improve important and negative aspects and support for tourism development by promoting positive and important variables. Involving the broader community furthermore ensures fewer problems and greater community acceptance than if only the business community or others outside the community are involved (Dimmock & Tiyce, 2001). The latter also emphasises the importance of determining the perceptions of those residents who visit the urban (national) park in their area and those who do not. This is especially important since Streuders (2008) found that not all communities understand the relevance of a national park and that without the input of the community, park management will not be aware of community needs or knowledge how to fulfil them.

Table 1 shows previous studies regarding residents' perceptions of urban and national parks. From these studies, the following findings are evident: 1) results show that parks in general have positive economic spin-offs and benefits to communities; 2) in most cases, communities are positive about national parks; and 3) communities also indicated that national parks contribute to their quality of life. Only one other study has focused on an urban national park in South Africa, namely the Wilderness National Park by Mouton (2010). The results indicated that this urban national park has a greater social impact than an economic impact. However, the study did not focus specifically on the perceptions of residents who visit the park and those who do not. The latter is, however, vital seeing as the role and contribution of local communities in managing national parks have changed significantly over the past few years. This has been brought about by a paradigm shift at South African National Parks from traditionally being a pure conservation agency to a conservation agency striving to benefit and empower local communities (Venter, Naiman, Biggs & Pienaar, 2008).

Table 1: Previous research regarding urban parks

Authors	Aim	Findings
Thompson (2002)	Urban open space in the 21st century	Urban parks provide space for expression of diversity, both personal and cultural.
Chiesura (2004)	The role of urban parks for the sustainable city	Natural areas (urban parks) contribute to quality of life and provide important social and psychological benefits to human societies, which enrich human life. Urban parks also fulfil important immaterial and non-consumptive human needs.
Saayman and Saayman (2006)	The socio-economic impact of the Addo Elephant National Park, South Africa	The Park has a significant impact on employment, production and general income creation in the region. The Park has also led to the establishment of businesses.

Authors	Aim	Findings
Salazar and Menédez (2007)	To determine non-market benefits of an urban park: Does proximity matter?	Mean willingness to pay (WTP) is considerably higher for people who live closer to urban parks.
Gidlöf-Gunnarsson and Öhrström (2007)	Noise and well-being in urban residential environments: The potential role of perceived availability to nearby green areas	Availability to nearby green areas is important for residents' well-being and daily behaviour by reducing long-term noise annoyance and urban environments.
Tyrväinen, Mäkinen and Schipperijn (2007)	Tools for mapping social values of urban woodlands and other green areas	Important features of urban green areas and woodlands include: tranquillity, the feeling of being in a forest and naturalness.
Ferreira (2008)	Socio-economic impact of the Karoo National Park, South Africa	The Park has a very positive impact on the quality of life of the residents of Beaufort-West. It has a greater social impact than an economic impact.
Mouton (2010)	Socio-economic impact of an urban national park: The case of the Wilderness National Park, South Africa	The results indicated that the Wilderness National Park has a greater social impact than an economic impact on the community of Wilderness.
Lo and Jim (2010)	Willingness of residents to pay and motives for conservation of urban green spaces in the compact city of Hong Kong	An urban park is an important place to visit family members and children as well as to exercise and breathe clean air. Eighty percent indicated a willingness to pay for recovering possible loss of urban green spaces.
Oberholzer (2010)	Socio-economic impact of the Tsitsikamma National Park, South Africa	Indicated that the Park has a positive environmental, social and economic impact.

3. Method of Research

As this was quantitative research, a structured questionnaire was used to collect the data. This section describes the questionnaire, the sampling method and survey, and the statistical analysis.

3.1 The Questionnaire

The questionnaire used for this research was adapted from previous work done by Slabbert, Saayman and Kruger (2009) regarding the social impact of the Kruger National Park in 2009. The questionnaire consisted of four sections. Section A captured demographic information (gender, home language, age, suburb, occupation and level of education), while Sections B and C measured the overall impact of the TMNP on the community and residents' visitation to the Park and travel motivations. Section D determined the social impacts of the TMNP on the surrounding communities. This section included 42 social impact items that were measured on a five-point Likert scale of agreement (1 = totally disagree and 5 = totally agree).

3.2 Sampling Method and Survey

In order to achieve the above mentioned aim, a survey was conducted regarding the social impact of the Park on neighbouring communities. Questionnaires from previous studies on social impacts conducted at Karoo and Kruger National Parks were modified to fit the description of TMNP. The social impact questionnaire collected demographic data, determined the overall impact of the TMNP on the communities, the local peoples' visitations to the Park as well as specific social impacts. Simple random sampling was chosen. The questionnaires were distributed by fieldworkers in surrounding communities within a radius of 1km from the Park. The following communities (towns) formed part of the research: Cape Town, Fish Hoek, Simon's Town, Constantia, Tokai, Muizenberg and Kalk Bay. Altogether, 551 questionnaires were successfully completed.

As shown in Table 2, the majority of respondents were female (63%), English speaking (78%), with an average age of 37 years, in sales positions (27%), well qualified (42%) and have lived in the community for an average of 19 years. Residents in the surrounding area feel that the TMNP has a positive impact on the community and their overall quality of life and that the Park is an asset to the community. However, the majority of residents had not visited the Park during the last year due to time constraints. This is confirmed by Slabbert, Saayman and Kruger (2009) and Ferreira (2008) in two separate studies regarding the socio-economic impact of the Kruger and Karoo National Parks, respectively. Those who had visited the Park did so to relax, to experience the outdoors, and for family recreation. This is also confirmed by research by Lee, Lee and Wicks (2004) and Yoon and Uysal (2005).

Table 2: Profile of TMNP communities

CATEGORY	Profile
Gender	Female (63%); Male (47%)
Home language	English (78%); Afrikaans (14%)
Age	Average: 37.3 years
Occupation	Sales personnel (27%); Self-employed (18%); Managerial (14%)
Level of education	Qualified (42%)
Impact personal quality of life	Positive impact (67%)
Impact on community	Positive impact (66%)
Park as asset to community	Yes (89%)
Visitation to Park 2009/10	No (52%)
Reasons for not visiting	Did not have time (64%)
Length of stay in community	19 years average
Reasons for visiting Park	To relax (81%); Experience outdoors (73%); Family recreation (70%)

3.3 Statistical Analysis

Microsoft© Excel© was used to capture data and SPSS (SPSS Inc, 2010) was used to analyse them. The analysis was conducted in two stages. Firstly, two principal axis factor analysis, using an Oblimin rotation with Kaiser normalisation, was performed on 13 travel motives items and the 42 social impact items to explain the variance-covariance structure of a set of variables through a few linear combinations of these variables. The Kaiser-Meyer-Olkin measure of sampling adequacy was also used to determine whether the covariance matrix was suitable for factor analysis. Kaiser's criteria for the extraction of all factors with eigenvalues larger than one were used because they were considered to explain a significant amount of variation in the data. In addition, all items with a factor loading above 0.3 were considered as contributing to a factor, and all with loadings lower than 0.3 as not correlating significantly with this factor (Steyn, 2000). In addition, any item that cross-loaded onto two factors with factor loadings greater than 0.3 was categorised in the factor where interpretability was best. A reliability coefficient (Cronbach's alpha) was computed for each factor to estimate the internal consistency of each factor. All factors with a reliability coefficient above 0.6 were considered as acceptable in this study. The average inter-item correlations were also computed as another measure of reliability – these, according to Clark and Watson (1995), should lie between 0.15 and 0.55.

Secondly, independent *t*-tests were used to investigate any significant differences between residents who visit TMNP and those who do not as well as between male and female respondents and home language. The study examined whether there were statistically significant differences between the two groups. Effect sizes were used to further identify any significant differences between the clusters. The purpose of effect size is to establish whether any differences exist between the groups; in this case, in which groups the mean values of the motives and social impact factors had the smallest or largest effect. Cohen (1988), Ellis and Steyn (2003) and Steyn (2009) offer the following guidelines for the interpretation of the effect sizes: small effect: d= 0.2, medium effect: d=0.5 and large effect: d=0.8. The results of the statistical analyses are discussed in the next section.

4. Results

This section is divided into two first results pertaining to the two factor analyses and second results regarding the *t*-tests.

4.1 Results from the Factor Analyses

Using an Oblimin rotation with Kaiser normalisation, the pattern matrix of the principal component factor analyses identified three motivational factors and eight social impacts (see Tables 3 and 4). These factors were labelled according to similar characteristics and accounted respectively for 58% (in both cases) of the total variance. All had relatively high reliability coefficients, ranging from 0.707 (the lowest) to 0.804 (the highest) for the motivational factors and 0.598 (the lowest) to 0.854 (the highest) for the social impact factors. Moreover, all items loaded on a factor with a loading greater than 0.3, and the relatively high factor loadings indicated a reasonably high correlation between the factors and their component items. The Kaiser-Meyer-Olkin measures of sampling adequacy of 0.88 for both the motivational factors and social impact factors also indicated that patterns of correlation were relatively compact and yielded distinct and reliable factors (Field, 2005). Bartlett's test of sphericity also reached statistical significance ($p < 0.001$), supporting the factorability of the correlation matrix (Pallant, 2007).

The results of the factor analysis regarding the travel motives are presented in Table 3.

Table 3: Factor analysis on travel motives

Factors	Attributes	Cronbach's alpha	Mean value
1) Park activities	To participate in adventure activities; To walk my dog; To do hiking trails; To do hiking trails; For photographic reasons; It is a spiritual experience	0.707	3.1
2) Family togetherness	For family recreation (to be with family); For the benefit of my children; To spend time with my friends; To learn about nature in general; To explore the mountain	0.804	3.6
3) Escape	To relax; To get away from my routine; To experience the outdoors	0.747	3.8

From the factor analysis, regarding travel motives, three factors were identified, namely *park activities*, *family togetherness* and *escape*. The factor with the highest mean value was *escape* (mean value 3.8) which was followed by *family togetherness* (mean value 3.6). This showing that most people visit Table Mountain National Parks to relax and spend time with their family. Various researchers, such as Fodness (1994), Uysal, McDonald and Martin (1994), Awaritefe (2004), Kruger and Saayman (2009), and Meng and Uysal (2008) confirm that escape and family togetherness are important travel motives for visitors to natural areas such as national parks.

For the factor analysis on the social impact of the park, eight factors were identified as listed in Table 4. The factor that had the highest mean value (3.8) and therefore seen as most important is *increased knowledge*. This was followed by *economic value* with a mean value of 3.7. *Social impact* and *environmental value* were also seen as important, with a mean value of 3.6 respectively.

Table 4: Factor analysis of social impacts

Factors	Attributes	Cronbach's Alpha	Mean Value
1) Economic Value	Number of tourists visiting has increased; Tourism and business development in the area are promoted; Money that tourists spend stimulate the economy; Opportunities for local businesses have increased; Roads and public facilities are better maintained.	0.803	3.7

Factors	Attributes	Cronbach's Alpha	Mean Value
2) Aspects of conflict	The Park has affected community negatively by relocating people for conservation purposes; The Park is insensitive towards the community concerning the removal of alien animal species	0.716	2.9
3) Social issues	The number of people in the area has increased; Prices of some goods and services increased; Crime levels have increased; Property value have increased; Traffic has increased; More investors are focusing on development in the area;	0.753	3.6
4) Information dissemination	Residents do not get sufficient information regarding activities in the Park; Residents do not get the opportunity to benefit economically from projects in the Park;	0.598	3.5
5) Environmental Value	The appearance of the area improved; Natural resources are conserved; The range of things to do in the area has increased; Employment in the area have increased; The natural environment has been sustained; Facilities have improved	0.818	3.6
6) Land control	There is not enough control over who enters the Park. Insufficient control causes veldfires; The Park needs to be fenced	0.632	3.5
7) Opportunities	The park gives opportunities to the community to use its natural resources e.g. Fishing and collecting fire wood; Funding for community activities has increased; Social and moral values have improved	0.643	3.2
8) Increased knowledge	Improves my knowledge with regards to the environment; Community has the opportunity to visit a national park; The TMNP is accessible for us as members of the community; Cape Town's reputation and tourism destination is enhanced; Increased my interest in activities such as hiking, birding	0.854	3.8

To be able to determine if there were any differences between residents who visit the Park as opposed to those who do not as well as whether there are differences in terms of gender and home language, regarding how they perceive the social impact of the Park, a t-test was conducted as discussed in the next section.

4.2 T-test Results

A t-test on three different groupings was carried out: female and male, English and Afrikaans, and yes or no if they had visited the Park.

From Table 5 it is clear that there were significant differences between male and female residents regarding the travel motives, *park activities* ($p = 0.01$) and *family togetherness* ($p = 0.02$); although the effect size differences are small. The results revealed that for male respondents *park activities* and *family togetherness* were seen as more important than for female respondents. Therefore, the findings suggest that gender has an impact on residents' travel motives to TMNP. No significant difference was found regarding gender for social impacts. This shows that male and female experience the social impact of the park similarly.

Table 5: T-test results of differences between female and male visitors pertaining to travel motives and social impacts

	Female			Male					
	Mean	**Std. dev.**	**N**	**Mean**	**Std. dev.**	**N**	**t-value**	**p**	**Effect size**
Travel motives									
Park activities	2.90	1.02	174	3.16	1.00	292	−2.59	**0.01***	**0.25****
Family recreation	3.50	0.92	174	3.70	1.00	292	−2.27	**0.02***	**0.21****
Escape	3.70	0.86	178	3.82	0.84	305	−1.29	0.20	0.12
Social impacts									
Economic issues	3.70	0.63	185	3.67	0.65	317	0.40	0.69	0.04
Conflicting issues	2.88	0.86	174	3.00	0.92	300	−1.49	0.14	0.14
Social issues	3.61	0.57	186	3.63	0.65	316	−0.36	0.72	0.03
Information dissemination	3.46	0.72	177	3.53	0.71	301	−1.01	0.31	0.10
Environmental importance	3.69	0.62	187	3.62	0.69	317	1.17	0.24	0.1
Land control	3.40	0.70	180	3.50	0.72	308	−1.57	0.12	0.15
Opportunities	3.16	0.64	182	3.19	0.69	310	−0.36	0.72	0.03
Increased knowledge	3.80	0.66	182	3.86	0.70	307	−0.92	0.36	0.08

*** Statistically significant difference: $p \leq 0.05$**

Effect sizes: ** small effect: d=0.2; ***medium effect: d=0.5 and ****large effect: d=0.8

In Table 6, the perceptions of English- and Afrikaans-speaking respondents were compared. No significant differences were found for travel motives to the Park as well as how they perceive the social impact of the Park on the local community. Therefore, the results indicate that "language" is not a contributing factor in determining the perceptions of residents regarding the social impact of the TMNP.

Table 6: T-test results of differences between English- and Afrikaans-speaking visitors pertaining to social impact and travel motives

	English			Afrikaans					
	Mean	**Std. dev.**	**N**	**Mean**	**Std. dev.**	**N**	**t-value**	**p**	**Effect size**
Travel motives									
Park activities	3.07	1.03	365	3.2	0.96	66	−0.96	0.34	0.13
Family recreation	3.61	0.93	376	3.73	0.87	70	−0.98	0.33	0.13
Escape	3.81	0.85	380	3.88	0.74	68	−0.62	0.53	0.08
Social impacts									
Economic issues	3.69	0.62	394	3.68	0.69	70	0.2	0.84	0.02
Conflicting issues	2.97	0.9	373	2.99	0.89	66	−0.21	0.84	0.03
Social issues	3.63	0.59	394	3.71	0.72	70	−1.06	0.29	0.12
Information dissemination	3.51	0.68	378	3.55	0.82	66	−0.42	0.67	0.05
Environmental importance	3.68	0.62	394	3.72	0.63	71	−0.51	0.61	0.06
Land control	3.47	0.7	385	3.56	0.75	67	−0.88	0.38	0.11
Opportunities	3.18	0.65	388	3.29	0.7	68	−1.33	0.18	0.17

*** Statistically significant difference: $p \leq 0.05$**
Effect sizes: ** small effect: d=0.2; ***medium effect: d=0.5 and ****large effect: d=0.8

The aim of Table 7 was to determine if there are any significant differences (t-value 0.00) between respondents who visited the Park compared to residents who did not visit the Park, regarding the social impact thereof and travel motives. Fifty-two percent of the respondents indicated that they had never visited the Park before and 48% did visit the Park. From the results, it was found that there is a significant difference regarding the travel motive *escape* ($p < 0.05$). Therefore, the analysis shows that people who visited the Park do feel stronger about *escape* as travel motive than those who did not visit the Park. A significant difference was furthermore found for the following social impact factors: *environmental importance* and *increased knowledge* ($p < 0.05$ in both cases). The results revealed that residents who visit the Park do feel that the Park has an impact regarding environmental education and to increase the knowledge of visitors about conservation and wildlife tourism.

Table 7: T-test results of differences between visitors who indicated that they visited the Park (yes) and those who did not (no) visit the Park pertaining to their travel motives and social impact

	Not visited the park			**Visited the park**					
	Mean	**Std. dev.**	**N**	**Mean**	**Std. dev.**	**N**	**t-value**	**p**	**Effect size**
Travel motives									
Park activities	2.96	1.00	212	3.14	0.99	236	−1.88	*0.06*	0.18
Family recreation	3.56	0.84	220	3.70	0.93	241	−1.77	*0.07*	0.16
Escape	3.58	0.86	219	3.96	0.78	242	−5.03	**0.00***	**0.45****
Social impacts									
Economic issues	3.67	0.66	239	3.72	0.59	238	−0.76	0.45	0.07
Conflicting issues	3.01	0.87	224	2.88	0.93	229	1.52	0.13	0.14
Social issues	3.62	0.65	239	3.63	0.56	240	−0.09	0.93	0.01
Information dissemination	3.48	0.73	224	3.50	0.71	232	−0.25	0.81	0.02
Environmental importance	3.55	0.69	239	3.77	0.58	243	−3.90	**0.00***	**0.33****
Land control	3.48	0.70	230	3.47	0.71	236	0.15	0.88	0.01
Opportunities	3.19	0.68	231	3.16	0.64	236	0.52	0.60	0.05
Increased knowledge	3.76	0.72	233	3.93	0.62	234	−2.82	**0.01***	**0.24****

*** Statistically significant difference: $p \leq 0.05$**
Effect sizes: ** small effect: d=0.2; ***medium effect: d=0.5 and ****large effect: d=0.8

5. Findings and Implications

This following findings and implications can be drawn from this research:

First, this research confirms that *escape* is the most common motive to visit national parks or nature-based products in South Africa, thereby confirming research by Fodness (1994), Uysal, McDonald and Martin (1994), Awaritefe (2004), Kruger and Saayman (2009), and Meng and Uysal (2008). The two remaining factors were *family togetherness* and *park activities*. This implies that TMNP can be successfully branded as the ideal get-away destination for residents and tourists alike. It also highlighted the importance of activities especially for an urban park. The challenge for this national park is to offer the right activities in a city such as Cape Town. Marketing messages should focus on the distinct experiences and activities offered by this urban national park.

Second, the research reveals several social impacts due to the park namely (in order of importance): *increased knowledge, economic value, social issues, environmental value, information dissemination, land control, opportunities* and *aspects of conflict*. These impacts have far reaching implications confirming the importance of an urban park and the role that it plays in the country where it is located. These impacts should also be used in emphasising the important role that urban parks play. This can be compared to an iceberg where the focus is generally on what people can see above the surface, for example infrastructure development, economic benefits and conservation, while a large number of positive aspects and impacts are not always visible, such as increased knowledge.

Third, communities do not only play an important role, but those who have visited the Park are more positive. These visitors can also be regarded as ambassadors of the Park. The implication is that South African National Parks (SANParks) needs to provide opportunities for residents to visit the Park and this can be achieved by means of season tickets and community days. This may help educate residents about the Park to understand the value and importance of conservation as well as to increase residents' knowledge pertaining to nature. This is confirmed by Baba Dioum, a noted African ecologist, who asserted "in the end we will conserve only what we love, we will love only what we understand and we will understand only what we are taught".

Fourth, residents who visit the Park see the Park as a place to escape to from their everyday routine. The implication is that SANParks can use this in marketing the Park, to residents, as a place to get away from your daily routine and work-related problems. Once again the type and variety of activities are important. The positive side is that residents do not have to travel far and they can visit frequently.

Lastly, this research confirms that a national park and specifically an urban park have a positive impact on the community and their overall quality of life and that the Park is an asset to the community. This is confirmed by Mouton (2010), Slabbert et al. (2009) and Oberholzer (2010) who did research on the socio-economic impact of national parks in South Africa. This also confirms the findings by Chiesura (2004) who added that an urban park contributes to quality of life, and provides important social and psychological benefits to human societies, which enrich human life. The implication is that SANParks can use this information to market urban national parks as areas that contribute to people's (residents') quality of life. This finding also implies that national parks offer more than just being areas of conservation but can also serve as a sanctuary for people to relax and improve their quality of life. This is an aspect that requires more work as well as acknowledgment from park authorities, wherethe focus is sometimes too much on the conservation value of parks, forgetting the various other benefits. This can also strengthen the argument for the development of future urban national parks. From a sustainability point of view, this finding is important, because an important sustainability outcome is an improved quality of life. The results also show that residents who visit the Park do feel that the Park has an impact regarding environmental education and to increase the knowledge of visitors about conservation and wildlife tourism which confirms the findings by Gobster (2001).

6. Conclusions

The aim of the present research was to determine whether the social impact of an urban national park (in this case TMNP) is perceived differently by residents who had visited the Park (48%) and those who have not (52%), and if these residents differ in terms of gender and home language. Eight factors were identified regarding the social impact of the TMNP, namely *park activities*, *family togetherness*, *escape*, *environmental value*, *land control*, *opportunities* and *increased knowledge*. The factors with the highest mean values were *escape* and *increased knowledge*. Therefore, one can state that the biggest social contribution that the TMNP has on the local communities around the Park is the fact that it provides residents with a place where they can escape to from their everyday lives and it provides an opportunity to increase their knowledge of conservation, wildlife and marine environments. The only significant difference found between residents who visit the Park and those who do not, is their perception of the social impact of the TMNP regarding the factors *environmental value* and *increased knowledge*. Therefore, one can conclude by saying that residents who visit the Park have a greater understanding regarding the importance that the urban parks play in environmental education and conservation. This can be because they have seen with their own eyes what the value of conservation is as opposed to those who had no first-hand experience of the Park. It is recommended that continued research is required to gain a greater understanding of especially the social impact aspects of not only urban national parks but other nature-based destination that are not always visible.

References

Adams, W.M. & Hutton, J. (2007). Review – people, parks and poverty: political ecology and biodiversity conservation. *Conservation and Society*, 5(2):147–183.

Akama, J. S & Kieti, D. (2007). Tourism and socio-economic development in developing countries: A case study of Mombasa Resort in Kenya. *Journal of SustainableTtourism*, 15(6):735–748.

Awaritefe, O.D. (2004). Motivation and other considerations in tourist destination choice: A case study of Nigeria. *Tourism Geographies*, 6:303–330.

Bixler, R.D. & Floyd, M.F. (1997). Nature is scary, disgusting and uncomfortable. *Environmental Behaviour,* 29:443–467.

Blamey, R.K. (2001). Principles of ecotourism. In Weaver, D.B. (ed). *The encyclopedia of ecotourism* (pp. 5–22). New York: CABI.

Bryant, M.M. (2006). Urban landscape conservation and the role of ecological greenways at local and metropolitan scales. *Landscape and Urban Planning,* 76(2006):23–44.

Chiesura, A. (2004). The role of urban parks for the sustainable city. *Landscape and Urban Planning,* 68(2004):129–138.

Clark, L. A. & Watson, D. (1995). Constructing validity: Basic issues in objective scale development. *Psychology Assessment, 7*(3):309–319.

Cohen, J. (1988). *Statistical power analysis for the behavioral sciences* (2nd ed.). Hillsdale, NJ: Erlbaum.

Cook, R.A., Yale, L.J. & Marqua, J.J. (2010). *Tourism: the business of travel* (4th ed.). Upper Saddle River, N.J.: Pearson Education.

Dimmock, K. & Tiyce, M. (2001). Festivals and events: Celebrating special interest tourism. In Douglas, N., Douglas, N & Derrett, R. (eds.), *Special interest tourism: Context and cases* (pp355–383). Brisbane, John Wiley & Sons Australia, Ltd.

Ellis, S.M. & Steyn, H.S. (2003). Practical significance (effect sizes) versus or in combination with statistical significance (p values). *Management Dynamics,* 12(1): 51–53.

Ferreira, M. (2008). *The socio-economic impact of tourism in the Klein Karoo National Park.* Unpublished Masters Dissertation. Potchefstroom: North West University, South Africa.

Field, A. (2003). *Designing a questionnaire.* [Online] Available at: http://www.statisticshell.com/docs/designing_questionnaires.pdf. Accessed 18 November 2013.

Field, A. (2005). *Discovering Statistics Using SPSS* (2nd edition). Thousand Oaks: SAGE.

Fodness, D. (1994). Measuring tourist motivation. *Annals of Tourism Research,* 21: 555–581.

Frank, L.D., Engelke, P.O. & Schmid, T.L. (2003). *Health and community design: the impact of the built environment on physical activity.* Washington, DC: Island Press.

Fredline, E. & Faulkner, B. (2001). Residents' reactions to the staging of major motor sport events within their communities: A cluster analysis. *Event Management*, 7(2):115–125.

Fredline, L., Jago, L. & Deery, M. (2003). The development of a generic scale to measure the social impact of events. *Event Management*, 8(1):23–37.

Geldenhuys, S. (2009). Ecotourism criteria and context. In Saayman, M. (ed.), *Ecotourism: Getting back to basics* (pp. 1–24). Potchefstroom: Platinum Press.

Gidlöf-Gunnarsson, A. & Öhrström, E. (2007). Noise and well-being in urban residential environments: the potential role of perceived availability to nearby green areas. *Landscape and Urban Planning*, 83(2007):115–126.

Gobster, P.H. (2001). Visions of nature: conflict and compatibility in urban park restoration. *Landscape and Urban Planning*, 56(2001):35–51.

Grahn, P. (1985). *Man's needs for urban parks, greenery and recreation*. Alnarp: Institute for landscape and planning, Swedish Agricultural University.

Grahn, P. & Stigsdotter, U.A. (2003). Landscape planning and stress. Urban forum. *Urban Green*, 2:1–8.

Grimm, N.B., Grove, J.M., Pickett, S.T.A. & Redman, C.L. (2000). Integrated approaches to long-term studies of urban ecological systems. *Bioscience*, 50(7):571–584.

Handy, S.L., Boarnet, M.G., Ewing, R. & Killingsworth, R.E. (2002). How the built environment affects physical activity. Views from urban planning. *American Journal of Prevention Medicine*, 23:64–73.

Hetzer, N.D. (1965). Environment, tourism, culture. Reprinted in *Ecosphere*, 1(2):1–3, July 1970.

Hoehner, C.M., Brennan Ramirez, L.K., Elliott, M.B., Handy, S.L. & Brownson, R.C. (2005). Perceived and objective environmental measures and physical activity among urban adults. *American Journal of Prevention Medicine*, 28:105–116.

Humpel, N., Owen, N., Iverson, D., Leslie, E. & Bauman, A. (2004). Perceived environment attributes, residential location, and walking for particular purposes. *American Journal of Prevention Medicine*, 26:119–125.

Jim. C.Y. (2004). Green-space preservation and allocation for sustainable greening of compact cities. *Cities*, 21:311–320.

Kaplan, R. (2001). The nature of the view from home. Psychological benefits. *Environmental Behaviour*, 33:507–542.

Kaplan, R. & Kaplan, S. (1989). *The experience of nature: a psychological perspective.* New York: Cambridge University Press.

Kruger, M. & Saayman, M. (2009). Travel motives to Kruger and Tsitsikamma National Parks: A comparative study. *South African Journal of Wildlife Research*, 40(1):93–102

Kuo, F.E. & Sullivan, W.C. (2001). Environment and crime in the inner city. Does vegetation reduce crime. *Environmental Behaviour,* 3(33):343–367.

Lee C.K., Lee Y.K. & Wicks B.E. (2004). Segmentation of festival motivation by nationality and satisfaction. *Tourism Management*, 25(1): 61–70.

Lo, A.Y. & Jim, C.Y. (2010). Willingness to pay and motives for conservation of urban green spaces in the compact city of Hong Kong. *Urban Forestry and Urban Greening*, 9(2010):113–120.

Mehta, J.N. & Heinen, J.T. (2001). Does community-based conservation shape favourable attitudes among locals? An empirical study of Nepal. *Environmental Management,* 28(2):165–177.

Meng, F. & Uysal, M. (2008). Effects of gender differences on perceptions of destinations attributes, motivations and travel values: An examination of a nature-based resort destination. *Journal of Sustainable Tourism,* 16:445–466.

Miller, R.W. (1997). Urban forestry: planning and managing urban greenspaces (2nd ed.). Englewood Cliffs, NJ: Prentice Hall.

Mouton, M. (2010). *Socio-economic impact of an urban park: The case of Wilderness National Park.* Potchefstroom: North-West University. (Unpublished Dissertation – MCom).

Oberholzer, S. (2010). *The socio-economic impact of the Tsitsikamma National Park.* Potchefstroom: North-West University. (Unpublished Dissertation – MCom).

Oberholzer, S., Saayman, S., Saayman, A. & Slabbert, E. (2009). The socio-economic impact of Africa's oldest marine park. Paper presented at the 6th *International Coastal and Marine Tourism Congress (CMT),* 24–25 June. Nelson Mandela University: Port Elizabeth, South Africa.

Office of Environment and Heritage. (2011). *The different types of parks and reserves.* http:www.environment.nsw.gov.au/parktypes/typesofparks.htm Date of access: 15 July 2011.

Page, S.J. & Connell, J. (2009). *Tourism: a modern synthesis* (3rd ed.). Hampshire: South-Western Cenegage Learning.

Pallant, J. (2007). *SPSS Survival Manual: A Step-by-step Guide to Data Analysis using SPSS Version 15.* (3rd ed.). New York: McGraw-Hill.

Ratz, T. (2000). Residents' perception of the socio-cultural impact of tourism at Lake Balaton, Hungary? In Richards, G. & Hall, D. (eds.), Tourism and sustainable community development (pp. 36–47). London/New York: Routledge.

Saayman, M. & Saayman, A. (2006). Creating a framework to assess the economic contribution of National Parks in South Africa: The case of the Addo Elephant National Park. *Tourism Economics*, 12(4):619–633.

Saayman, M. & Saayman, A. (2009). Why travel motivations and socio-demographics matter in managing a national park. *Koedoe*, 51(1):381.

Saayman, M., Van Der Merwe, P., Saayman, A., & Mouton, M. E. (2009). The socio-economic impact of an urban park: the case of Wilderness National Park. *Journal of Policy Research in Tourism, Leisure and Events*, 1(3):247–264.

Salazar, S. & Menendez, L.C. (2007). Estimating the non-market benefits of an urban park: Does proximity matter? *Land Use Policy*, 24(2007):296–305.

SANParks (South African National Parks). (2009). *Table Mountain National Park.* http://www.sanparks.org/parks/table_mountain/ Date of access: 15 April 2009.

SANParks (South African National Parks). (2010). *Table Mountain National Park.* http://www.sanparks.org/parks/table_mountain/ Date of access: 25 June 2010.

Slabbert, E., Saayman, M. & Kruger, M. (2009). *Social impact of Kruger National Park.* Unpublished report. Potchefstroom: Institute for Tourism and Leisure Studies.

SPSS Inc. (2010). *SPSS® 16.0 for Windows, Release 16.0.0.* Chicago: SPSS Inc.

Standish, B., Boting, A., Van Zyl, H., Leiman, T. & Trupie, J. (2004). *The Economic contribution of Table Mountain National Park.* Cape Town: The Graduate School of Business, University of Cape Town.

Steyn, H.S. (2000). Practical significance of the difference in means. *South African Journal of Industrial Psychology*, 26(3):1–3.

Steyn, H.S. (2009). *Manual: Effect size indices and practical significance.* [Online] Available at: http://www.nwu.ac.za/p-statcs/index.html. Accessed 25 May 2013

Streuders, C. (2008). *Communication efficacy of South African National Parks: a case study of Karoo National Park.* Unpublished Masters Dissertation. Potchefstroom: North West University.

Taplin, D.H., Scheld, S. & Low, S.M. (2002). Rapid ethnographic assessment in urban parks: a case study of Independence National Historic Park. *Human Organisation*, 61(1):80–93.

Taylor, C.N., Gough, J., Warren, J & McClintock, W. (1999). Social and economic impacts of Kahurangi National Park: Science for conservation. Wellington, New Zealand: *Department of Conservation.*

Thompson, C.W. (2002). Urban open spaces in the 21st century. *Landscape and Urban Planning*, 60(2002):59–72.

Tyrväinen, L., Mäkinen, K., & Schipperijn, J. (2007). Tools for mapping social values of urban woodlands and other green areas. *Landscape and Urban Planning*, 79(1):5–19.

Uysal, M., McDonald, C.D. & Martin, B.S. (1994). Australian visitors to US national parks and natural areas. *International Journal of Contemporary Hospitality Management*, 6:18–24.

Van der Merwe, P. & Saayman, M. (2004). *Managing game farms from a tourism perspective.* Potchefstroom: Institute for Tourism and Leisure Studies.

Venter, F.J., Naiman, R.J., Biggs, H.C. & Pienaar, D.J. (2008). The evolution of conservation management philosophy: Science, environmental change and social adjustments in Kruger National Park. *Ecosystems*, 11:173–192.

Walker, J.R. & Walker, J.T. (2011). *Tourism: Concepts and practices.* Upper Saddle River, NJ: Prentice Hall.

West, P., Igoe, J. & Brockington, D. (2006). Parks and people: the social impact of protected areas. *The Annual Review of Anthropology*, 35:14.1–14:27.

Yoon, Y. & Uysal, M. (2005). An examination of the affects of motivation and satisfaction on destination loyalty: A structural model. *Tourism Management*, 26(1), 45–46.

Jan Velvin and Kristian Bjørnstad

Multifunctionality, Tourism and Protected Areas

1. Introduction

Over the last few decades rural areas in Europe have faced several challenges. The use of rural areas for the production of food and fibers is still important in terms of the area used, however, in terms of employment and in percentage of the economy, the overall trend is that the productive function of rural areas has lost much of its previous dominant position. At the same time, there are indications of rural land users and other stakeholders assigning a higher value to the consumption, largely through tourism, and the heritage of rural areas. Multifunctionality is now a key term in relation to this transition and the driving forces behind it has been suggested as being related to agricultural overcapacity, the emergence of market-driven amenity uses and a growing societal awareness of sustainability and preservation issues (Holmes, 2006). In Norway, this is visible through an increased focus on nature-based tourism in rural areas and in the planning and implementation of new protected areas (Direktoratet for Naturforvaltning, 2006).

Rural areas and rural policies in Norway have had a prominent place among policy makers and in the general population. This has provided rural areas with strong government services and one of the most subsidised agricultural sectors in the world. However, this has not reversed the trend of migration from rural to urban areas and the reduction of active farmers. To meet these challenges, there is an increased focus on diversifying the rural economy by developing a variety of commercial activities. As in other parts of Europe, tourism plays an increasingly important part in this process. In some rural regions in the southern part of Norway, for example in Hallingdal, tourism already employs one third of the population directly and indirectly (Dybedal, 2005).

As framework conditions for traditional farming and forestry activities are changing, the development of nature-based tourism through the sustainable use of protected areas can represent a resource for local land users and rural communities. Especially as studies have shown that designating protected areas, primarily as national parks, increases the flow of tourists (Reinius and Fredman, 2007). Although seemingly a resource, there have been few fruitful connections between area protection processes and the development of commercial activities among land users in Norway. In fact, area protection is considered something to avoid among land users and there are often conflicts when centralised protection processes are initiated. Subsequently, there is a great need for more insight into the development of commercial activities in relation to protected areas. This is especially urgent since there appears to be a paradigm shift in European area protection policies, with more integrated and dynamic approaches aimed at reconciling nature conservation with the needs of local people in a context of sustainable development (Mose and Weixlbaumer, 2007).

Challenging more traditional area protection approaches, there are a number of new and more integrated area protection initiatives emerging in Norway, such as regional nature parks and biosphere reserves. In continental Europe, these have long played a key role in the protection and development of rural areas. These new initiatives can be characterised as being multifunctional since they target the multiple functions of rural areas in more integrated and participatory ways, and often paying special attention to bottom up processes, learning, and competence development among a variety of stakeholders (Mose and Weixlbaumer, 2002, Hammer et. al., 2007, Mose and Schröder, 2012). Although not common in Norway, elements of such approaches are making its way into area protection processes in rural areas. This makes it urgent to study this in more detail.

The primary objective of this chapter is to contribute insight and knowledge about the background for development of nature-based tourism activities among actors in and around protected areas in peripheral rural regions in Norway. Our research questions are: (1) what factors influence farm diversification strategies among land owners in protected areas in Norway? (2) How and to what extent can more integrative area protection approaches influence farm diversification strategies?

2. The Ecostrategic Framework

We draw our theoretical framework on current literature within the interdisciplinary fields of rural studies, tourism, and environmental studies and have employed the ecostrategic framework as a strategic approach to integrate and to advance the discussion on the variation of views and practices regarding the utilisation of nature and landscapes (Sandell 2005, Zacharisson et al., 2006). Each of the four ecostrategic approaches involve a set of consequences in terms of land use, development, participation and democracy.

The ecostrategic framework takes its departure point in the tension between the adaptation of human activities in a context of local resource use in natural and cultural landscapes, and the large scale domination of nature and landscapes through functional specialisation. This can also be described as a tension between the perspectives of multifunctional land use, and large scale one-dimensional specialisation. In Figure 1, this is illustrated through the horizontal line. In agricultural terms at the farm level, this could describe the tension between farm strategies emphasising specialisation and mechanisation into one activity, versus a diversification strategy emphasising multiple farm activities based on local resources.

Figure 1: The conceptual framework of ecostrategies (Sandell 2005, Zachrisson et al., 2006)

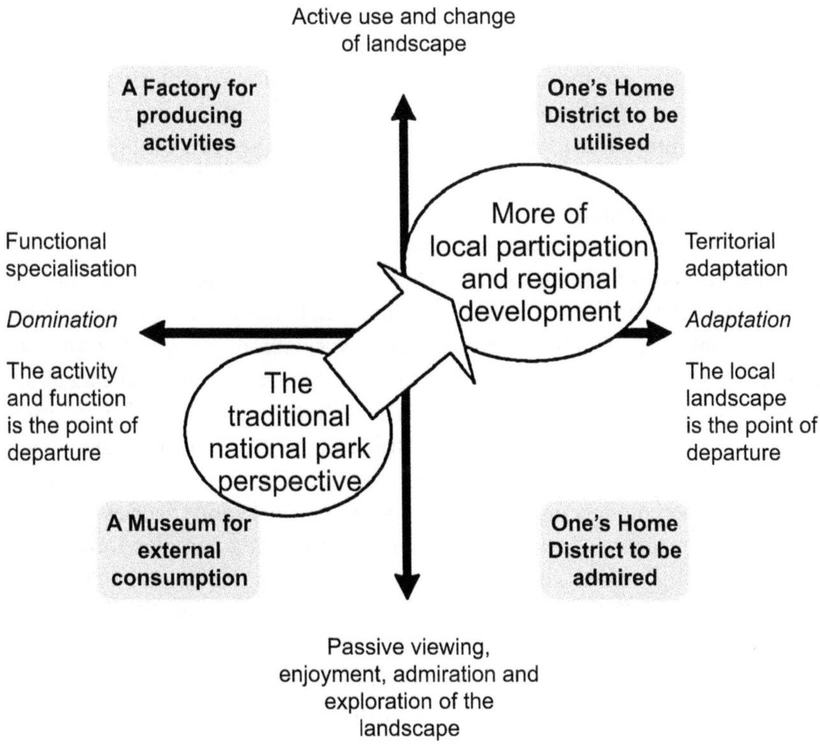

The vertical line describes the tension between active use versus passive viewing of nature and landscapes. Here, we can introduce and connect questions regarding area protection since we can place the traditional national park in the bottom left corner of the figure. In traditional conservation approaches the goal was, and often still is, to freeze or to conserve a certain area or landscape for biodiversity protection and as a "museum" for external consumption. Meanwhile, the top left corner can serve to illustrate a recent development in regards to tourism activities in and around some protected areas, namely for the area to serve as a "factory" for producing activities, for example when external actors develop motorised activities in and around national parks.

What characterises new area protection approaches, the so called integrated or dynamic-innovative approaches, is the inclusion of local residents in the protection process and the development of multiple services based on local resources.

The broad arrow across Figure 1 illustrates the relevant ecostrategies implied in a shift from more traditional to new area protection approaches. It is important to note that the conceptual framework of ecostrategies are not fixed in any way; they are more about directions for various adaptation strategies guiding human relations with nature and landscape. In this regard, it is also possible that more traditional conservation approaches can contain dynamic elements.

To view multifunctionality and ecostrategies in relation to sustainability is a topic on its own and not as apparent as it might seem, since something can be multifunctional yet not sustainable. However, it could be argued that multiple small-scale activities based on local resources can better form the potential basis for sustainable development than large-scale specialised use. Another question regarding sustainability is which dimensions of sustainability will potentially be supported by the various ecostrategies. A traditional national park might contribute to ecological sustainability, while contributing less to social and economic sustainability. In a discussion on sustainability, it is important then to use and develop models that can synthesise various data and theories about development alternatives in order to be able to better view complex interactions between environmental, social and economic systems (Patterson et. al., 2004). In articulating and conceptualising a new integrated area protection paradigm this will be an important task.

The strength of the ecostrategic framework is that it creates a conceptual framework for connecting notions of multifunctionality with various area protection approaches and land use strategies. It also illustrates the change in area protection approaches in a very visible way. The ecostrategic framework can also be useful for illustrating the tensions between various views and practices, something which can inform an analysis of conflicts in area protection processes. A further refinement of the ecostrategic framework can involve more knowledge about how to connect the strategic with the operative levels.

3. Multifunctionality and Diversification

In agricultural and rural research the development of complementary commercial activities among land users have made use of the closely related terms "multifunctionality", "farm diversification, "part-time farming", or "pluriactivity" (Evans and Ilbery, 1993, Busby and Rendle, 2000, Reinius and Fredman 2007, McGehee et al., 2007, Barbieri et al., 2008, Wilson 2008, Barbieri and Mahoney, 2009, Brandth and Haugen, 2011). Multifunctional activities, or diversification activities, are often referred to as commercial activities on a farm or forest property which is run in addition to, or beside traditional agriculture or forestry activities, where one or more of the farm's resources are included. Another common usage of the term multifunctionality is in relation to the various functions of agriculture, "agriculture is multifunctional producing not only food but also sustaining rural landscapes, protecting biodiversity, generating employment and contributing to the viability of rural areas" (Potter and Burney, 2002 p. 35). They also demonstrate the interconnectedness of these functions, as "sustaining rural landscapes" is of crucial importance for a tourism industry, dependent on such landscapes as part of customer satisfaction.

Multifunctional land use practices have been present in various degrees among land users for a long time, but it is unclear exactly when the concept of multifunctionality was introduced in policy discourses[1]. Multifunctionality is a complex and problematic concept lacking a common definition. There are few commonly agreed upon definitions of the concept, and it tends to take on different meanings among different persons and groups of people (De Vries, 2000). However, the concept often refers to the various commodity (food and fibers) and non-commodity outputs that are jointly produced by agriculture (OECD, 2001). Non-commodity outputs are also described as "public goods", which can be rural landscapes, biodiversity, rural viability and food safety. These goods, both private and public, can be important in the wider rural development process in that they might play a part in attracting new rural settlers, in strengthening rural tourism and recreation activities, and in the marketing of regional products (Bryden and Dawe, 1998). It is also important to note that that there are conflicts over the goods among groups of actors benefiting from goods that the rural region provides (Terluin, 2003). An example of this can be in the implementation of area protection initiatives, where various stakeholders tend to display differences, for example, byvaluing different goods, such as rural viability or biodiversity.

1 The notion of "multifunctionality" was mentioned in 1994 in the Uruguay Round agreements on agricultural trade (Losch 2004).

In European agricultural and rural research, the concept of multifunctionality has received increased attention. However, a major problem with the concept is that it has not been developed in relation to local and regional contexts (Vandermeulen, et. al., 2006, Renting et al., 2008). Furthermore, there have been few attempts in Norway and other countries, at studying multifunctionality in connection with area protection initiatives (Wilson, 2008, Falleth and Hovik, 2009,), although a fruitful approach for connecting types of land use with various types of area protection approaches is the conceptual framework of "ecostrategies" (Sandell, 2005 and Zacharisson et. al., 2006).

Diversification strategies and commercial activities among land users

Nature-based tourism is an increasingly important activity in rural areas (Aas et al., 2003). It can be defined as a form of tourism where the natural environment is the most important attraction, and where tourist activities are situated and conducted in close proximity of this environment (Saarinen, 2000). In addition to being a resource for the individual land user, the sustainable development of small-scale tourism activities (e.g., cabin rentals, animal safaris) can contribute to the development of local communities (McGehee and Kyungmi, 2004, McGehee et al., 2007, Marsden and Sonnino, 2008, Phillip et al., 2010).

Rural policies and integrative area protection initiatives promoting more nature-based tourism and other multifunctional activities can seem to be a resource for farmers, foresters and other rural actors by responding better to the need for a more consumer- or consumption-oriented countryside focused on leisure and nature-based tourism (Marsden, 1999). At a higher level, these activities can be important in wider rural development processes by playing a part in attracting new rural settlers, in strengthening rural tourism and recreation activities as well as the marketing of regional products (Bryden and Dawe 1998). The integrated development of such activities is a well known strategy (Jenkins et. al. 1998, Marcoullier et. al., 1998, Hall, 2000, Hall and Mueller, 2004).

Although tourist activities and other commercial farm activities are increasing in importance in Norway, they are still considered by land users to be an additional activity on farming and forestry properties beside traditional commodity production (Kjesbu et al., 2007, Velvin and Krogh, 2012). In the process of moving from an understanding in which traditional agricultural activities are prioritised to one in which new and additional activities and roles are valued in a multifunctional approach (for example as host and tourist guide) it can be a difficult adjustment for many farmers in Norway. If the area protection processes do not offer the right mix of facilitation, competence development and the development of new products (as in more integrative approaches), it can prove to be a challenge for farmers to engage in these processes.

It is evident that the factors influencing the development of diversification activities among land users and the wider rural development processes are not well understood, especially in connection with area protection initiatives. Research shows that the creation of networks is an important factor in farmer diversification (Mathijs, 2003, Jongeneel et al., 2005), making it relevant to examine the existence and the qualities of networks among local land users. This is particularly important as networks and partnerships are important elements in providing the context for interaction, learning and co-operation among actors developing sustainable tourism products (Saxena, 2005). Networks can also form the basis for a type of social and cultural entrepreneurship, which in turn can be of benefit for community development (Ateljevic and Doorne, 2000, Clemetsen et al., 2007, Dacin et. al., 2010).

4. Integrative and Dynamic Approaches Towards Area Protection

Area protection policies and initiatives are currently a central part of rural areas. Up until the last few decades, a fairly static preservation approach towards area protection was dominant in Europe and especially in the Nordic countries. Traditionally, the goal of area protection was primarily designed to protect certain species in specific areas, with as little human interference as possible. Implementation and planning of such initiatives has been characterised as a static-conservative approach being mainly a top-down process paying little attention to the needs and wishes of the local population (Mose and Weixlbaumer, 2002, Hammer et. al., 2007).

In contrast to much of the Norwegian research on protected areas, other European researchers are increasingly focusing on integrated models of area protection, which is visible in initiatives such as regional nature parks and biosphere reserves in central and southern Europe. In more integrative, or dynamic-innovative approaches towards area protection, the goal is to integrate functions, such as the preservation of biodiversity, recreation, regional development and environmental education. This is possible with an increased use of zoning (protection, buffer and development zones) and in a context of sustainability and low intensive land use (Mose and Weixlbaumer 2002, Hammer, 2007, Hammer et. al., 2007). An integrated area protection approach will pay closer attention to stakeholder participation, collective learning and network development in the area protection process. This can be of potential importance for the diversification activities of farm entrepreneurs and local land users, as research shows that participation in networks of agricultural and non-agricultural organisations increases the likelihood of diversification activities (Mathijs, 2003, Jongeneel et al., 2005). There are currently a number of more integrative and dynamic approaches towards area protection in Norway; however, these processes are not prioritised among researchers.

Although with this background and the research gap that currently exist in Norway, the objective of this study is to seek answers to our second research question, on how more integrative approaches, or elements of such approaches, might influence farm diversification activities among land users in and around protected areas.

5. A New Approach to Area Protection in Rural Norway

In Norway, conflicts often arise in connection with area protection initiatives. These conflicts are generally attributed to centralised decision making with little real participation by local actors (Velvin et. al., 2010). A contributing factor is that there is little research on processes, tools and appropriate network formations in regards to area protection processes. This can be attributed to the fact that research and development of protected areas in Norway has traditionally been influenced by natural sciences, and less by social and economic perspectives (Sjeggedal, 2007, Velvin et. al., 2010, Velvin and Krogh, 2012).

In recent years more integrated perspectives seem, however, to be emerging in both the research and development of protected areas in Norway, visible in development projects connected with national parks. These projects conclude that there is already a considerable amount of commercial activity in and around Norwegian protected areas (Aas et.al., 2003, Heiberg etal., 2005). However, the main challenge is finding a balance between the profitable, and the ecologically and socially sustainability.

There have also been recent experiments where the development of commercial activities and local participation waere incorporated into the area protection process. The case of Junkerdal National Park has shown that such approaches strengthened stakeholder relationships and reduced conflict (Bay-Larsen and Sanderson, 2005). Another example of a more integrative approach in Norway is the establishment of a regional park in the municipality of Aurland, in connection with a large state run conservation project (Clemetsen et al., 2007) and Trillemarka Nature Reserve in Buskerud County (Velvin et. al., 2010, Velvin and Krogh, 2012). This provides some indications that an exploration of more integrative area protection approaches can be a relevant and necessary task in a Norwegian context. Early results show that more integrative approaches can reduce conflict and help facilitate cooperation among farmers and other stakeholders in area protection processes. There is still much research needed on the additional values and the concrete effects on the diversification strategies of farmers in and around protected areas.

6. Concluding Remarks

Conservation processes in rural areas in Norway appear to start changing from a static approach with exclusive focus on the ecological qualities to a more dynamic form, where the local people in the communities that are affected by protected area processes are included in discussions where the various sustainability elements (economic, social and ecological) are seen as equivalent approaches. The insights gained from process such as in the Trillemarka Nature Reserve invite us to realise the full intent of the law (Naturmangfoldloven); coordination of conservation, and sustainable use (Velvin et. al., 2010). This multiple approach objective will require a balanced co-management in processes with real local participation, and where knowledge linked to the scientific field of ecosystems is combined with cultural issues, social sciences and economic studies of the local use of the protected area. The result could be more network cooperation, or differentiated zoning of areas related to conservation and sustainable use, which also maintains opportunities for sustainable economic development (Sandell, 2005, Hammer, 2007, Velvin et. al., 2010)

We have seen that more integrative and dynamic area protection approaches can help reduce conflict and help facilitate cooperation. However, if business development is a goal in connection with protected areas in Norway, our study show that it is not sufficient to have protective provisions that facilitate business activity and increased level of zoning. It also requires the development and change of network facilitation and competence development to achieve a sustainable use proficiently. Such competence will have to include customer-related expertise and the development of identity in relation to industrial activities.

Further investigation of multifunctionality, tourism and area protection is recommended. It should be examined in a broader research perspective with more focus on additional values and concrete effects. This may also include social and cultural values involving farmers from different rural areas involved in various area protection initiatives.

References

Aas, Ø., Vistad O. I., Dervo B., Eide N.E., Kaltenborn B.P., Haaland H., Andersen O., (2003). Management and Use of National Parks in Mountains: – International Experiences on Policy and Management of Human Activities in National Parks; – Mapping of Business Activities in the Three Norwegian National Parks Dovrefjell-Sunndalsfjella; Femundsmarka and Reisa; – a Literature Study of Ecological, Cultural and Social Effects from Tourism in National Parks. Lillehammer: *NINA Fagrapport*, 72. 83pp.

Ateljevic, I. and Doorne, S. (2000). 'Staying Within the Fence': Lifestyle Entrepreneurship in Tourism. *Journal of Sustainable Tourism,* 8, 378–392.

Barbieri, C. & Mahoney, E. (2009). Why is diversification an attractive farm adjustment strategy? Insights from Texas farmers and ranchers. *Journal of Rural Studies,* 25, 58–66.

Barbieri, C., Mahoney, E. and Butler, L. (2008). Understanding the Nature and Extent of Farm and Ranch Diversification in North America. *Rural Sociology,* 73, 205–229.

Bay-Larsen, I. and Sanderson, H. (2005). Medvirkningsorientert partnerskap I verneplanprosesser – eksempelet Junkerdal Nasjonalpark. *Tidsskriftet Utmark,* vol. 1. (www.utmark.org)

Busby, G. and Rendle, S. (2000). The transition from tourism on farms to farm tourism. *Tourism Management,* 21, 635–642.

Brandth, B. and Haugen, M. S. 2011. Farm diversification into tourism – Implications for social identity? *Journal of Rural Studies,* 27, 35–44.

Bryden, J. and Dawe, S P. (1998). Development Strategies for Remote Rural Regions: What do we know so far? *OECD International Conference on Remote Rural Areas – Developing Through Natural and Cultural Assets.* Spain: November 6–8 1998.

Clemetsen, M., Krogh, E., Moulton, M. (2007). Creativity on a Landscape Scale – The Role of Social and Cultural Entrepreneurship in Sustainable Community Development. A Study from the Fjord landscape of Western Norway. In proceedings from workshop, *Cost Action C27, Sustainable Development for Minor Deprived Urban Communities,* Evora: 23rd to 25th April 2007.

Dacin, P. A., Dacin, M. T. and Matear, M. (2010). Social entrepreneurship: why we don't need a new theory and how we move forward from here. *The Academy of Management Perspectives,* 24, 37–57.

De Vries, B. (2000). *Multifunctional Agriculture in the International Context: A Review.* Lewiston: The Land Stewardship Project.

Direktoratet for Naturforvaltning (2006). *Rapport til Miljøverndepartementet om handlingsplan for bærekraftig bruk, forvaltning og skjøtsel av verneområder.* Trondheim: Direktoratet for Naturforvaltning.

Dybedal, P. (2005). Ringvirkninger av reiseliv I Buskerud, Telemark og Vestfold. Oslo: *TØI rapport 780/2005.*

Evans, N. J., Ilbery, B. W. (1993). The pluractivity, part-time farming, and farm diversification debate. *Environment and Planning A.* 25.7. 945–99.

Falleth, E. I. and Hovik, S. (2009). Local government and nature conservation in norway: decentralisation as a strategy in environmental policy. *Local Environment: The International Journal Of Justice And Sustainability,* 14, 221–231.

Hammer, T. (2007). Protected Areas and Regional Development: Conflicts and Opportunities. In: Mose, I. (Ed.): *Protected areas and regional development. Towards a new model for the 21 century?* Aldershot, 21–38.

Hammer, T., Mose, I., Siegriest, D., Weixlbaumer, N. (2007). Protected Areas and Regional Development: Towards a new model for the 21 century? In: Mose, I. (Ed.): *Protected areas and regional development. Towards a new model for the 21 century?* Aldershot, 233–246.

Hall, M. (2000). *Tourism Planning.* Harlow: Prentice Hall.

Hall, M. and Müeller, D. (2004). Introduction: Second Homes Curse or Blessing? Revisited.

In Hall and Müeller (Eds): *Tourism, Mobility and Second Homes-Between elite landscape and common ground* Ontario: Channel View, 3–14.

Heiberg, M.M., Cristensen, H.M. and Aas, Ø. (2005). Turisme i verneområder. Forprosjekt. Trondheim: *NINA Rapport,* 87.

Holmes, J. (2006). Impulses towards a multifunctional transition in rural Australia: Gaps in the research agenda. *Journal of Rural Studies,* 22, (2), pp. 142–160.

Jenkins, J. (1998). The restructuring of rural economies: Rural tourism and recreation as a government response. In R. Butler, C.M. Hall and J. Jenkins (eds). *Tourism and recreation in rural areas,* New York: John Wiley & Sons Ltd, 43–68.

Jongeneel, R., Polman, N., Slangen, L. (2005). Why are farmers going multifunctional? *XI International Congress of the EAAE,'The Future of Rural Europe in the Global Agri-Food System',* Copenhagen, 24–27.

Losch, B (2004). Debating the multifunctionality of agriculture: from trade negotiations to development polities by the south. *Journal of Agrarian Change 4 (3), 336–360.*

Kjesbu, E., Hegrenes, A., Rye Paulsen, S.K., Sjelmo, O., Stokke, K. (2007). *Tilleggsnæringer og annen næringsaktivitet på yrkesmessig drevne bruk*. Notat. NILF.

Marcoullier, D.W., Green, G. P., Deller, S.C., Sumathi, N.R. (1998). *Recreational Homes and Regional development: A case study from Upper Great Lakes States*. Madison: University of Cooperative Extension Publications.

Marsden, T., (1999). Rural Futures: the consumption countryside and its regulation. *Sociologia Ruralis*, 39 (4), 501–526.

Marsden, T. and Sonnino, R. (2008). Rural development and the regional state: Denying multifunctional agriculture in the UK. *Journal of Rural Studies*, 24, 422–431.

Mathijs, E. (2003) Social capital and farmers' willingness to adopt countryside stewardship schemes. *Outlook on Agriculture*, 32 (1):13–16.

McGehee, N. G. and Kyungmi, K. (2004). Motivation for Agri-Tourism Entrepreneurship. *Journal of Travel Research*, 43, 161–170.

McGehee, N. G., Kim, K. and Jennings, G. R. (2007). Gender and motivation for agri-tourism entrepreneurship. *Tourism Management*, 28 (1), 280–289.

Mose, I., Weixlbaumer, N. (2002). Naturschutz: Großschutzgebiete und Regionalentwicklung (= Naturschutz und Freizeitgesellschaft, Bd. 5). Aachen.

Mose, I. and Schröder, I. (2012). *Tourismus in Europa*, Oldenburg: Oldenburgische Volkszeitung.

Mose, I., Weixlbaumer, N. (2007). A new paradigm for Protected Areas in Europe?. In: Mose, I. (Ed.): *Protected areas and regional development. Towards a new model for the 21 century?* Aldershot, 3–20.

OECD (2001). *Multifunctionality: towards an analytical framework*, Paris. OECD.

Patterson T., Cousins, K., Kraev, E. (2004). Integrating environmental, social and economic systems: a dynamic model of tourism in Dominica. *Ecological Modelling*, 175:. 121–136.

Phillip, S., Hunter, C. and Blackstock, K. (2010). A typology for defining agritourism. *Tourism Management*, 31, 754–758.

Potter, c. and Burney, J. (2002). Agricultural multifunctionality in the WTO-legitimate non-trade concern or disguised protectionism? *Journal of rural studies*, 18, 35–47.

Renting, H., Oostindie, H., Laurent, C., Brunori, G., Barjolle, D., Moxnes Jervell, A., Granberg, L., Heinonen, M. (2008). Multifunctionality of agricultural activities, changing rural identities and new institutional arrangements.

International Journal of Agricultural Resources Governance and Ecology, 7 (4–5), 361–385.

Reinius, S.W. and Fredman, P. (2007). Protected areas as attractions. *Annals of Tourism Research,* 34(4), 839 - 854.

Saarinen, J. (2000). Snapshots of the history of nature-based tourism in Finland. In: Saarinen, J. (Ed.) *The Transformation of a Tourist Destination. Theory and Case Studies on the Production of Local Geographies in Tourism in Finish Lapland.* Nordia Geographical Publications, 30 (1), pp. 95–105.

Sandell, K. (2005). *Reservatsdilemman eller utvecklingsområden? En syntes-diskussion och två fallstudier: Torneträsk biosfärsområde 1986–2004 och Södra Jämtlandsfjällen nationalparkförslag.* Umeå: FjällMistra, Fjällmistra rapport.

Sjeggedal, T. (2007). Verneplanprosesser – reelle konflikter eller institusjonelle forviklinger? *Tidsskriftet Utmark.* Vol. 1.

Saxena, G. (2005). Relationships, networks and the learning regions: a case evidence from the Peak District National Park. *Tourism Management,* 26:277–289.

Terluin, I.J., (2003). Differences in economic development in rural regions of advanced countries: an overview of and critical analysis of theories. *Journal of Rural Studies,* 19, 327–344.

Vandermeulen, V., Verspecht, A., Van Huylenbroeck, G., Meert, H., Boulanger, A., Van Hecke, E. (2006). The importance of the institutional environment on multifunctional farming systems in the peri-urban area of Brussels. *Land Use Policy,* 23, 486–501.

Velvin, J., Krogh, E., and Vedeld, P. O. (2010). Når enden er god er vel allting godt? Vern av Trillemarka-Rollagsfjell – styringsprosesser, deltakelse og politiske nettverk. *Utmark – tidskrift for utmarksforskning, 10*(1). (www.utmark.org)

Velvin, J., and Krogh, E. (2012). Naturoppfatning og turisme -implikasjoner for rural turistbasert næringsutvikling. *Utmark – tidskrift for utmarksforskning, 12*(2). (www.utmark.org)

Wilson, G. A. (2008). From 'Weak' to 'Strong' Multifunctionality: Conceptualising Farm-Level Multifunctional Transitional Pathways. *Journal of Rural Studies,* 24, 367–383.

Zacharisson, A., Sandell, K., Friedman, P., Eckerberg, K. (2006). Tourism and protected areas: motives, actors and processes. *International Journal of Biodiversity Science and Management,* (2) 350–358.

Jonathon Peter Spring

Caught in the Crossfire: Using Participant Observation to research Guided Tours

1. Introduction

Research into guided tours can provide a window into the human dimension of tourism. Guided tours are an opportunity for both the voices of the host community, guides and visitors to be heard and where the meanings that people attach to the places experienced can be explored. Participant observation is a research technique that can provide rich data for understanding tourism phenomena (Cole, 2005; KwaZulu-Natal, 2008, March), and has been used in researching guided tours (Benton, 2009; Holloway, 1981; Howard, Thwaites, & Smith, 2001; Lugosi &Bray, 2008; Seaton, 2002).

With academic research on human behaviour, the relationship between the researcher and the research participants is an essential factor. The issue of disclosure has ethical, ontological and epistemological implications about the nature of the data collected (Sieber, 2004; Van Deventer, 2009).

The scientific foundation of academic research posits a research norm where research participants should be unaware of their involvement in research lest that awareness creates a significant bias that would render the results meaningless. In tourism research, a preference for research based within a positivist paradigm, and a lack of debate about the applicability of other ontological and epistemological frameworks of research has been discerned (Pritchard & Morgan, 2007; Tribe & Airey, 2007).

Institution-based parameters governing research on human behaviour are a feature of many universities to protect all involved in the research community from harm. Participant observation can represent an intrusion into the working life of the guide and the private life of the visitor, and if there are no options for consent to participate, the data may be invalid due to ethical considerations.

Emergent researchers face the challenge of trying to devise methods that will aid them to uncover valid insights from their research aims. In trying to consider the use of participant observation in researching guiding tours three potential hurdles arise: a) making sense of the methodological implications of using participant observation in conjunction with full or non-disclosure; b) a procedural process through institutional review boards and ethics committees that appears to view the use of participation observation without full disclosure and full consent as ethically unacceptable; and c) the practical problems posed in trying to deploy either full disclosure and full consent or alternatively non-disclosure.

The author utilised participant observation involving full disclosure on learning during guided wildlife tours. The author's experience of mediating concerns about the validity of data collected within guided tours at scientific and ethical levels forms the basis of this chapter. This chapter attempts to make sense of these issues and the challenges of addressing both positions on participant observation in fashioning a methodology that can be employed in real world settings.

2. Participant Observation

Participant observation is a field research method derived from the ethnographic tradition (Atkinson & Hammersley, 1994). It is a technique that seeks to make sense of phenomena associated with the social and cultural practices of a society or community. The researcher records observations made while participating in the social group being studied (Plester, 2007). The extent of the participation can vary from actually doing the specific activity being observed or acting as an onlooker or audience (Atkinson & Hammersley, 1994; Czarniawska, 2004). There are attempts to differentiate what can be considered participant observation in relation to other modes of observation used in field research (Seaton, 2002; Stewart, Hayward, Devlin, & Kirby, 1998).

With its use as a research method in twentieth century anthropology, the researcher was often perceived by the research participants, as being 'visibly' different to the participants, either on grounds of gender, ethnicity or social background (Teshome-Bahiru & Negashe-Wossene, 2007). With the expansion of areas of accepted study in the sociological and anthropological fields in the social and work places of industrial and post industrial societies, the potential for the researcher to be anonymous increased (Weaver, 2002). A researcher using participant observation for research in tourist settings may be construed to be a tourist themselves by the research participants.

2.1 The Use of Participant Observation in Tourism Research and Researching Guided Tours

In a study of North American tourists in Europe, Hartmann (1988, p. 101) found that participant observation used in conjunction with interviews provided 'unexpected insights otherwise unobtainable by any of the other field research methods used'. The use of participant observation in tour guiding has been infrequent. Seaton (2002, p. 317) notes that 'given the possibilities of real-life observation in tourism research, it is surprising that there have been few participant observation studies of tourists'.

The author's research on the guided tours of Pacific Whale Foundation (PWF) in Hawai'i was part of an ongoing study to explore guided wildlife tours to establish the scenarios where visitors are most receptive to learning and to examine whether that learning is enhanced through visitor-guide interactions. Participant observation was combined with interviews with both visitors and guides. The observations informed the content of the semi-structured interviews.

Field research at PWF took place between 1st December 2009 and 7th January 2010 in Maui, Hawai'i (Table 1). In total the field research involved 31 boat-based snorkelling and whale-watch trips. Formal observations of naturalists occurred on 28 boat trips from 4th December 2009 to 4th January 2010. From these observations 12 visitors were invited to participate in interviews after the boat trips. The naturalists who were observed on the tour the visitor participated in were then interviewed. From October 2010 to April 2011, 14 follow up interviews occurred with visitors and guides who were initially interviewed in Maui.

Table 1: Pacific Whale Foundation trips participated in for study (1st December 2009 to 7th January 2010)

Boat and port (Lahaina = L & Ma'alaea = M	Snorkelling	Whale watching (sunrise)	Whale watching (day)	Observations	Guide interviews (follow up interviews)	Visitor interviews (Follow up interviews)
Discovery (L)		5	2	7	4 (2)	4 (3)
Explorer (L)		1		1	0	0
Intrigue (M)			9	8	3 (1)	2 (1)
Odyssey (M)	3		4	6	1	1
Quest (L)	1			1	0	0
Spirit (M)	1	4		4	4 (4)	4 (3)
Voyager (M)	1			1	1	1
Sub totals	6	10	15	28	12 (7)	12 (7)
Totals	**31**			**28**	**38**	

When deciding to employ participant observation in researching tourism phenomena, the researcher has to consider the notion of disclosure and its implications on the validity of data in respect to ethics, ontology and methodology.

Use of participant observation within tourism research in conjunction with full disclosure may have implications for those whose ontological perspective puts into question knowledge generated from the behaviour of informed participants. For use of participant observation using covert tactics from an ethics approach based on medical criteria or where colonisation may have been part of the political landscape, the lack of disclosure and or full consent are deviations that need to be negotiated and fully justified between the researcher, the faculty and the ethics body of the research institute (Fitzgerald, 2005; Hudson & Russell, 2009; Tolich, 2002). A discussion on disclosure can help highlight the specific problems and concerns raised about participant observation in conjunction with covert and full disclosure.

3. Disclosure and Participant Observation in Tourism Research

When researching guided tours the researcher has to decide to what extent the subjects are involved. Discussion on disclosure can relate to practical, methodological, and ethical issues. Practical issues may be cost and frequency of observations in relation to the nature of site, frequency of tours, and the extent to which guides or employees of an organisation involved in the research may recognise and become familiar with the researcher. Methodological issues may include disclosure and how its use or non-use may affect the findings. Ethical issues include the potential harm to the participants as well as the researcher and how to avoid or minimise the potential for risk. Overlap exists in these areas and can occur in scenarios where the study incorporates formal interviews as well as participant observation that necessitate a level of disclosure over and above that which was used with participant observation prior to the interviews. This can create a process where the decision not to inform the participants from the outset may, in itself, create a bias in the findings.

Participant observation strategies in tourism research have employed covert methods ranging from total deception where participants are unaware that research is taking place (Hartmann, 1988; Miller, Hudson, & Turner, 2005; Seaton, 2002) to situations where the participant may know that they are the subjects of the research but not know what specific part of their experience is the focus of the research (Holloway, 1981) to full disclosure (Maat, 2009). Research on guided tours outside the tourism discipline has also used full disclosure (Benton, 2009; Salazar, 2008).

Where there has been a level of disclosure provided, and the researcher's methodological standpoint may recognise their visibility during the data collection phase as an issue in relation to the implications of their visibility, the researcher may seek to employ tactics that will reduce their visibility, minimisng their own impact on the natural quality of the data being collected.

For research involving non-disclosure the researcher has to conceal their role as a researcher. This can affect data collection techniques such as questioning or note taking that may not appear to be natural behaviour within the context of the research setting (Ryan & Martin, 2001).

3.1 Implications of Disclosure and Consent on the Validity of Data and the Notion of Bias

Full disclosure by explicitly stating the aims of the research to participants can introduce an important potential bias that may shape results. That is, research subjects who understand they are being researched may shape their behaviour to what they perceive as desirable for the research. The researcher then has to determine the extent of the impact disclosure has on the participant and the influence this may have on the validity of the research findings (Plester, 2007).

It is a truism of research that 'the part' is often used to make inferences about or give insights to 'the whole'. In determining the validity of the part in being representative of the whole, the obtrusiveness of the techniques on gathering data is considered a key factor. The researcher as an 'outsider' may create a bias in the research where participants stop acting naturally. The participant may shape their behaviour to conform to what the participant perceives to be desirable in the context of being observed by an outsider (Herrera, 1999).

Determining the validity of data is a complex issue that is informed by the nature of the research, the ontological and epistemological framework of the research, and the discipline within which the study occurs. A key issue in the devising of the research methods for observing guided tours is to what extent would disclosure to the research participants (in this case guides and tourists) of the research affect their natural behaviour. The author decided to use full disclosure in a pragmatic manner that aligned to the schedule of the boat-based tours. Disclosure to the guide occurred prior to the trip while disclosure of the research to the visitors was made during the captain's formal welcome to the visitors as the boat left the harbour. Written consent of the participants in the interviews was obtained after the tour, but prior to the interviews. The methods utilised in the field work with Pacific Whale Foundation were based on the idea that the participants' awareness of the researcher's presence will have some impact on their behaviour (Table 2) but that over the course of the guided tour the presence of the whales and the total sum of staff visitor interaction would be more influential factors. The priority for the research was that there was a shared experience of a guided tour between the researcher and the participants from which pertinent phenomena could be explored.

Table 2: Disclosure of study to participants during observations on whale watching tours, Pacific Whale Foundation, Maui, Hawai'i

Captain's announcement to the visitors: Jonathon is all the way from New Zealand. He is doing a PhD Study on learning during guided tours and he will be with [name of guide involved in the observation] during the tour. He may ask one or two of you to participate in the research. If you want to know any more please do not hesitate to ask him. There is also a laminated information sheet on the study at the bar.

Sample of comments recorded by researcher	Persons involved	Site	Total passengers on board	Context
"a tip for your study; she's talking too much"	Visitor to researcher	PWF whale watching tour Lahaina Sunrise tour	56	Comment made in passing while I was on the top deck towards the stern of the vessel while the lead naturalist was on the microphone. Observation did not generate interviews.
Visitor informs me that they have visited Nelson and Christchurch in New Zealand	Visitor to researcher	PWF whale watching tour Ma'alaea day tour,	22	Comment made moments after the first pod of humpback whales were encountered. Observation generated interviews.
"So where are you getting your doctorate from?"	Visitor to researcher	PWF whale watching tour Ma'alaea day tour	88	Comment made while observing whales travelling. Observation did not generate interviews.
"boat coming out of harbour is going to hit the whale" (looks at me with concern)	Visitor to researcher	PWF whale watching tour Ma'alaea Sunrise tour	47	Comment made while observing whales close to harbour mouth where there were vessels waiting to enter and exit the harbour mouth. Observation generated interviews.
"What's your study in?"	Visitor to researcher	PWF whale watching tour Ma'alaea Sunrise tour	34	Question asked in final stages of the trip while still encountering humpback whales. Not long after researcher asked the visitor to participate in an interview.

3.2 Ontological and Epistemological Implications of Full Disclosure

All academic research is underpinned by the ontological and epistemological assumptions about the nature of knowledge and the appropriate methods by which to obtain that knowledge (Guba & Lincoln, 2005). The researcher's own understanding of the relationship between disclosure and ontological and epistemological nature of knowledge is sought.

The nature of how the researcher records their observations also then plays a role in the decision about disclosure. Note taking within sight of the participants may create questions in the mind of the visitors. These can be explained as travel diary entries (Seaton, 2002). Other tactics such as going to secluded spaces such as a toilet to surreptitiously write notes have also been used in tourism research.

While observing PWF boat-based guided tours, the author conspicuously wrote notes throughout the duration of the tour. This allowed him to record in real time what tourists were observing without having to disappear into a toilet or other discrete space to record. The author's note taking was noticed by visitors and may have had an impact on what they said to the guide the researcher accompanied as well as to other visitors (see Table 2).

Research strategies that emphasise reducing the visibility of the researcher are aligned to a positivist paradigm that implies that disclosure to the participant of the researcher's presence through informed consent or visible appearance will render the data unacceptably biased. An interpretivist or constructivist paradigm advocates the agency of the research subject in the development of knowledge and thus informed consent and the visibility of the researcher can contribute to the validity of the data (Guba & Lincoln, 2005). However the transient nature of observations in some research contexts may make it difficult to use a participant observation strategy that employs full disclosure and full consent (Lugosi, 2006).

A review of four tourism journals from their inception to 1996 noted a bias towards positivism and suggested a marginalisation of newer social science research developments within the field of tourism research (Tribe & Airey, 2007). For research based within a positivist paradigm any level of disclosure to the research subject could invalidate the data collected in terms of the research objectives (Calvey, 2008; Crow, Wiles, Heath, & Charles, 2006).

The positivist research norm is that a human subject is unaware that they are research participants in order to generate knowledge that can hold up to external and internal validity checks. They are research subjects rather than participants. Minimal disclosure is required so that their awareness of the research does not invalidate their contribution. The constructivist research norm holds that the research participants are active participants in creating the findings and so disclosure to the participants about the research is an important factor influencing their ability to contribute (Drury & Stott, 2001; Guba & Lincoln, 2005).

A possible implication of the role of institutional review boards and ethics committees in the research design process is that the expected procedure for all field work involving participant observation should be full disclosure. What is more important in the search of knowledge using participant observation; whether non-disclosure may result in some form of harm to the participant or whether full disclosure could affect the trustworthiness of the findings?

3.3 Implications of the Ethics of Research and the Research Institution on Participant Observation and Use of Non-disclosure

An important consideration in the use of participant observation is at what level i is necessary for the research participant to give their consent to be part of the research The relationship between the researcher and research participants is having a major bearing on what academic institutions view as appropriate areas of research and informed consent is a fundamental principle of ethical research (Haggerty, 2004; Hedgecoe, 2008).

The creation of institutional ethics committees has been an attempt to create a more independent approach to judging the appropriateness of research, the notion of risk and issues of informed consent and full disclosure. A 1988 government review of a medical study on cervical cancer framed the discourse on research ethics in New Zealand. The implications that this case had at an institutional level for all New Zealand academic research is the need to consider at an ethical level what impact the research may have on the overall well-being of the participants (Fitzgerald, 2005).

Aside from the researcher's considerations about the ethical implications of the research, institutional review boards and ethics committees have the potential to impose provisions and limits on the use of research instruments. With participant observation there can be an insistence that observation occurs in conjunction with full disclosure and/or full consent:

> Institutional Review Boards and ethics committees expect researchers to predict where and when issues of concealment will emerge, assess the implications of such concealment and construct bureaucratized forms of disclosure and consent that demonstrably negate any potential risk. It is assumed that concealment is thus avoidable because the contractual agreement between researchers and informants clearly define their mutual rights, roles, and obligations, and consensual participation is based on an appropriate understanding of the research (Lugosi, 2006, p. 545).

Formal ethics review processes create another temporal dimension to the research requirements for emergent researchers engaged in postgraduate research. The defined time limits of Masters and PhD research means that using participant observation with non-disclosure could create time delays that may not be possible for the student to negotiate within prescribed time limits of their study. A postgraduate student may believe that it is more expedient to utilise full disclosure with participant observation for time management reasons rather than giving due consideration to its impact on their research.

3.4 Participant Observation and the Notion of Risk

The discourse of risk in terms of academic research can be summarised as relating to misrepresenting, exploiting, harming or removing the anonymity of the research subjects and harm to the researcher (Crow et al., 2006; Hedgecoe, 2008; Sieber, 2004).

In the context of participant observation, the researcher's study may unwittingly misrepresent the research subject or their community in their interpretation of the data collected. The study may be exploitative in that the study has potentially developed opportunities to the researcher, in say, career advancement but little in the way of the participants. Harm to the participant can arise if they discover later that they are part of a study and then regret the behaviour observed. This is intensified, intentionally or otherwise if the participants' identity is revealed. With sociological studies within the researcher's own community or country the researcher's presence could have an impact on the participants or put the researcher at risk.

The transient nature of the tourist population at sites has been cited as a factor which differentiates research on tourists from other communities when it comes to ethical issues such as disclosure and informed consent (Hartmann, 1988). This does not account for the residents working or partaking in leisure activities in the research areas. The presence of residents in the form of the host community, guides or the tourism workers with tourists during guided tours has resulted in a dichotomy in the use of full or non-disclosure in research. For example, in Holloway's (1981) study of one day guided tours in London, full disclosure was made to the guides and drivers of the tours in conjunction with a request to participate in interviews after the observations. The tourists were unaware that they were participating in a research act.

In guided tours, the risks for the guide and other employees associated with the tour may relate to concerns about how the findings may affect their job security. The research may highlight deficiencies in their work performance, and this may be used by their employers against them.

One dimension of the temporal setting in respect to the visitor is that it occurs during their leisure time. A point made to the author by a staff member of his university's ethics review board in respect to the ethics of the research involving PWF tours was that the research was an intrusion into tourists' leisure sphere.

There are a range of potential risk factors that a researcher has to consider in the design of research involving participant observation and guided tours. The possible risk to participants who are working as guides is much easier to understand at a psychological level in relation to fear of losing their jobs. The potential psychological harm on a tourist caused by either their awareness or lack of awareness of participating in research during their leisure time is a much more difficult proposition to understand.

4. Conclusion

In order to progress tourism research it has been argued that "alternative perspectives are necessary if new insight into the nature of tourism is to be generated" (Faulkner & Russell, 2003, p. 205). As participant observation has been an integral research technique in anthropology and sociology, protocols have been established to reduce the bias that visibility could cause (Teshome-Bahiru & Negash-Wossene, 2007). How informed consent, or non-disclosure, affects the validity of data collected through participant observation has not been fully explored within the discipline of tourism. There has been a lack of debate in tourism studies about the utility of the visibility of the researcher in relation to participant observation. A tension exists in tourism research between employing research techniques that adhere to the positivist paradigm aligning research to institutional ethics guidelines that advocate for informed consent. Institutionalised ethics reviews act as a check to ensure no harm occurs to all involved in the research process but there is a perception that it imposes constraints on what can be legitimately researched. The practical realities of using participant observation in researching guided tours can make it difficult to fully adhere to either non-disclosure or full disclosure. It would have been very difficult for the author to have stayed totally anonymous to PWF staff while observing 31 of their tours over the period of five weeks.

Researchers may then rely on proven techniques when researching guided tours where a process has been identified or tradition established within tourism where certain tests can be implemented to test the validity of the data, and the ethicality of the instruments. This may result in a limited set of research techniques available to current and prospective tourism researchers and impact on the scope of tourism research. The ultimate decision for employing either disclosure or non-disclosure with participant observation rests with the researcher and their ability or inclination to devise a research method that does not intentionally result in some form of harm to the participant and does not affect the trustworthiness of the findings of their research.

References

Atkinson, P., & Hammersley, M. (1994). Ethnography and participant observation. In N. K. Denizen & Y. S. Lincoln (Eds.), *Handbook of qualitative research* (pp. 248–261). Thousand Oaks, CA: Sage.

Benton, G. M. (2009). From priniciple to practice: four conceptions of interpretation. *Journal of Interpretation Research, 14*(1), 7–31.

Calvey, D. (2008). The art and politics of covert research: Doing 'situated ethics' in the field. *Sociology, 42*(5), 905–108.

Cole, S. (2005). Action ethnography: Using participant observation. In B. W. Ritichie, P. M. Burns, & C. A. Palmer (Eds.), *Tourism research methods: Integrating theory with practice* (pp. 63–72). Wallingford, UK: CABI.

Crow, G., Wiles, R., Heath, S., & Charles, V. (2006). Research ethics and data quality: The implications of informed consent. *International Journal of Social Research Methodology, 9*(2), 83–95.

Czarniawska, B. (2004). *Narratives in social science research.* London: Sage.

Drury, J., & Stott, C. (2001). Bias as a research strategy in participant observation: The case of intergroup conflict. *field methods, 13*(1), 47–67.

Faulkner, B. & Russell, R. (2003). Chaos and complexity in tourism: In search of a new perspective. In L. Fredline, L. Jago & C. Cooper (Eds.). *Progressing tourism research – Bill Faulkner* (pp. 205–219). Clevedon, England: Channel View).

Fitzgerald, M. H. (2005). Punctuated equilibrium, moral panics and the ethics review process. *Journal of Academic Ethics, 2*, 315–338.

Guba, E. G., & Lincoln, Y. S. (2005). Paradigmatic controversies, contradictions and emerging confluences. In N. K. Denizen & Y. S. Lincoln (Eds.), *The Sage handbook of qualitative research* (3rd ed., pp. 191–216). Thousand Oaks, CA: Sage.

Haggerty, K. D. (2004). Ethics creep: Governing social science research in the name of ethics. *Qualitative Sociology, 27*(4), 391–414.

Hall, C. M., & Kearsley, G. (2001). *Tourism in New Zealand: An Introduction.* Melbourne, Vic, Australia: Oxford University Press.

Hartmann, R. (1988). Combining field methods in tourism research. *Annals of Tourism Research, 15,* 88–105.

Hedgecoe, A. (2008). Research ethics review and the sociological research relationship. *Sociology, 42*(5), 873–886.

Herrera, C. D. (1999). Two arguments for 'covert methods' in social research. *British Journal of Sociology, 50*(2), 331–343.

Holloway, J. C. (1981). The guided tour: A sociological approach. *Annals of Tourism Research, 3,* 377–402.

Howard, J., Thwaites, R., & Smith, B. (2001). Investigating the roles of the indigenous tour guide. *The Journal of Tourism Studies, 12*(2), 32–39.

Hudson, M. L., & Russell, K. (2009). The Treaty of Waitangi and research ethics in Aotearoa. *Journal of Bioethical inquiry, 6*(1), 61–68.

KwaZulu-Natal, T. (2008, March). *The strengths and weaknesses of qualitative research in understanding the behviour of tourism and travel consumers* (60). Retrieved from http://www.zulu.org.za/userfiles/1/file/ForIvestorandResearcher/Research/Occ%20PDFs/Occ%20papers%20PDFs/Occ%20paper%2060%20QualTourismResearch.pdf

Lugosi, P. (2006). Between overt and covert research: Concealment and disclosure in an ethnographic study of commercial hospitality. *Qualitative Inquiry, 12*(3), 541–561.

Lugosi, P., & Bray, J. (2008). Tour guding, organisational culture and learning: lessons from an entrepreneurial company. *International journal of tourism research, 10,* 467–479.

Maat, V. (2009). *The dynamics of a singles holiday: The transformative potential of short-term singles travel.* (MSc thesis). University of Wageningen, Wageningen.

Miller, G., Hudson, S., & Turner, R. (2005). Applying the mystery shopping technique: The case of Lunn Ploy. In B. W. Ritchie, P. Burns, & C. Palmer (Eds.), *Tourism research methods: Integrating theory with practice* (pp. 63–72). Wallingford, England: CABI.

Plester, B. (2007). *Laugh out loud: How organisational culture influences workplace behaviour*(PhD). Massey University, Albany, Auckland.

Pritchard, A., & Morgan, N. (2007). De-centring tourism's intellectual universe, or traversing the dialogue between change and tradition. In I. Ateljevic,

A. Prtichard, & N. Morgan (Eds.), *The critical turn in tourism studies: Innovative research methods* (pp. 11–28). Oxford, England: Elsevier.

Ryan, C., & Martin, A. (2001). Tourists and strippers: Liminal theater. *Annals of Tourism Research, 28*(1), 140–163.

Salazar, N. B. (2008). *Envisioning Eden: A glocal ethnography of tour guiding* (PhD dissertation). University of Pennsylvania, Philadelphia.)

Seaton, A. V. (2002). Observing conducted tours: the ethnographic context in tourist research. *Journal of Vacation Marketing, 8*(4), 309–319.

Sieber, J. E. (2004). Using our best judgement in conducting human research. *Ethics & Behavior, 14*(4), 297–304.

Stewart, E. J., Hayward, B. M., Devlin, P. J., & Kirby, V. G. (1998). The "place" of interpretation: a new approach to the evaluation of interpretation. *Tourism Management, 19*(3), 257–266.

Teshome-Bahiru, W., & Negashe-Wossene, J. (2007). Anthropological field researches in the 21st Century: Scope, challenges and ethics. *Anthropologist Special volume, 3*, 7–18.

Tolich, M. (2002). Pakeha "paralysis": Cultural safety for those researching the general population of Aotearoa. *Social Policy Journal of New Zealand, 19*, 164–178.

Tribe, J., & Airey, D. (2007). A review of tourism research. In J. Tribe & D. Airey (Eds.), *Developments in tourism research* (pp. 3–14). Oxford, England: Elsevier.

Van Deventer, J. P. (2009). Ethical considerations during human centred overt and covert research. *Quality and Quantity, 43*, 45–57.

Weaver, T. (2002). Everett C. Hughes: Urban sociology, social problems and ethics. In T. Weaver (Ed.), *The dynamics of applied anthropolgy in the twentieth century* (pp. 45–58). Retrieved from http://www.sfa.net/malinowski/monograph/malinowskimonograph.html. Retrieved 15 December 2010

Xiaoshu Sun and Michael Lück

The Internet presence of whale and dolphin watch operators in New Zealand in terms of their ecotourism attributes: A content analysis

1. Introduction

Awareness of environmental issues has become increasingly significant worldwide, and since the late 1980s, ecotourism has developed remarkably, particularly attributed to the public becoming tired of the crowds, awakened to the evidence of pollution, and in search of something new (Fennell, 2002; Blamey, 1995). Thus, an alternative to unregulated mass tourism was sought in ecotourism. The term "ecotourism" has gained substantial attention and interest from governments, industries, academics and communities (Blamey, 1997), as its development provides a tourism experience which is not harmful to the local community and natural environments. With "ecotourism" becoming a catchword in the tourism industry since the 1990s (Honey, 1999), it has today become an integral part of economic strategy and developmental policy in a great number of regions and countries. Ecotourism is one of the fastest growing sectors in the tourism industry (Wood et al., 2002), and as Western (1993) noted, it "has surged through the travel and conservation world like a tsunami" (p. 7). According to Weaver (1998), ecotourism is a concept which has evolved over the last few decades in the conservation community, and with local people in natural areas. Ecotourism has brought various opportunities to achieve conservation goals and improve the well-being of local communities, promising a significant win-win situation in these areas (Drumm & Moore, 2002).

Wildlife tourism is a significant form of ecotourism, which has been widely accepted as a growing industry, with particularly high growth rates in the whale watching tourism sector (Higginbottom, 2004). However, marine experts have expressed concerns around the whale watching operations' potential negative impacts on the marine life. Therefore, a number of researchers have investigated these aspects in studies on whale watching tourism in many areas around the world (Higham & Lück, 2007).

2. Whale Watching

Whale watching is generally accepted as feasible, sustainable ecotourism and a more advantageous exploitation of marine mammals than lethally harvesting for products (International Fund for Animal Welfare, 1995). According to Woods-Ballard et al. (2003), whale watching is defined by the IWC (1994) as "any commercial enterprise which provides for the public to see cetaceans in their natural habitat" (p. 33). Hoyt (2000) adds more detail, by defining whale watching as "tours by boat, air or from land, formal or informal, with at least some commercial aspect, to see, swim with, and/or listen to any of the some 83 species of whales, dolphins and porpoises" (p. 1). Nowadays, whale and dolphin watching activities are taking place on every continent and in countries as diverse as South Africa, Argentina, Norway, Japan, New Zealand and Tonga (Orams, 2000). Whale watching is claimed to be one of the most successful forms of ecotourism in the world, which provides economic revenues, conservation, educational, and research benefits. It thus has the potential to provide maximum benefits to both the tourists and the targeted species (Hoyt, 2007).

However, a number of researchers have warned that the close approach taken by tourism boats for watching, or swimming with dolphins and whales, is altering these marine animals' behaviour (Phillips & Baird, 1993). It has been suggested that these tourism activities could be detrimental to the marine wildlife (Beach & Weinrich, 1989). Another pressing problem worldwide in recent years is the issue of marine pollution. A significant proportion of this waste is derived from the irresponsible practices of whale watching tourists both on a personal level and from poor tourism practices, while in and around the marine environment (Lück & Higham, 2007). In addition, whale watching activities could have a considerable impact through noise pollution on marine mammals that are exposed due to the frequent interactions between the tourism platforms and targeted marine species (Lusseau, 2007). Moreover, boat collisions are significant physical threats that pose a serious threat to cetaceans' lives (Vanderlaan & Taggart, 2007). According to Laist et al. (2001), a number of marine mammals have been injured or killed as a result of collisions with whale watching vessels in the past, especially in areas with a high intensity of whale watching traffic.

3. Whale Watching in New Zealand

Since the first commercial whale watching operation began at Kaikoura in 1987, New Zealand has witnessed a significant growth in whale watching visitor participation numbers (Lück & Altobelli, 2009). As Constantine (1999) asserted, the opportunities to watch cetaceans is a great motivation for a large number of tourists to come to New Zealand. In 2008, the domestic and international tourists made more than 546,445 visits for whale watching in New Zealand.

The estimated total expenditure (both direct and indirect) on whale watching tourism in 2008 was almost $81 million (O'Connor et al., 2009).

Lück (2009) noted that management of marine mammals in New Zealand is based on two main parts of legislation: the Marine Mammals Protection Act (1978) and the Marine Mammals Protection Regulations (1992). According to the MMPA, all commercial companies operating within conservation estate in New Zealand are required to have permits which are issued by the Department of Conservation. The requirements to hold a permit from the Department of Conservation are multi-faceted, and applications are assessed carefully. The MMPR control the requirements with regard to permits, suspension, revocation, restriction or amendment of permits, behaviour around marine mammals, and other miscellaneous provisions.

4. Methodology

As an information-intensive industry, tourism gains significant synergies from the use of the internet (Garin-Munoz & Perez-Amaral, 2011). Websites have become a suitable medium for tourism operators to conduct marketing and promotion for their businesses, as a tourism website enables users to quickly access information of interest. Content analysis is an approach for the analysis of documents and texts, which seek to quantify content in terms of predetermined categories and in a systematic and replicable matter (Bryman, 2012). There are currently 72 operators providing various whale watching activities in New Zealand (personal communication, Department of Conservation), most of whom (64 operators) have their own websites. These 64 websites were the objects of research in this study. For this research, a manifest content analysis of the websites of all whale watching operators in New Zealand was conducted (Babbie, 1995). There are three different approaches in the applications of content analysis, including: conventional, directed or summative (Weber, 1990). The conventional approach is deemed a suitable application of content analysis in this research. In a conventional content analysis, the coding categories are directly derived from the text information, which is commonly utilized in research designed or aimed at describing a phenomenon. By utilizing this approach, the researcher could avoid introducing preconceived ideas and categories, and allow the categories and names of categories to come directly from the research data.

Data were analysed by using thematic analysis, which is the one of the most common methods in qualitative analysis (Mutch, 2005). The process of thematic analysis consists of nine stages (Figure 1). It is a strict step-by-step inductive procedure (Braun & Clarke, 2006).

Figure 1: Nine stages of thematic analysis in the research

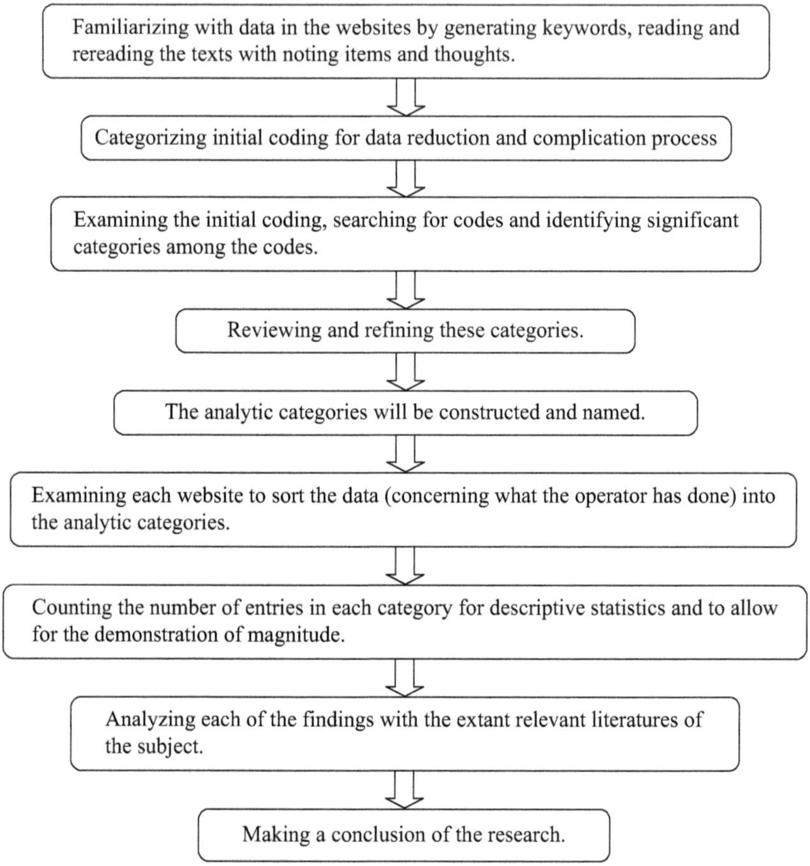

Source: After Braun & Clarke (2006)

5. Results and Discussion

By studying the sixty-four New Zealand whale watching operators' websites, a variety of contributions for ecotourism could be elicited. Accordingly, the data from the websites were gathered and converted into text, and codes were inductively identified in the data, then transformed into the categorical themes. After examining and refining these themes about the contributions for ecotourism, nineteen analytic categories were identified and established. Then, each of the websites was examined to sort the data concerned with what the operator has presented on their website, into the nineteen analytic categories (Table 1).

Table 1: Ecotourism Attributes of New Zealand Whale Watching Operators, Presented on their Websites

Contributions / Whale Watching Operators	Eoo-operator	Qualmark	Purchasing and Operation					Staff Management		Education		Waste Control			Conservation Work				
			Purchasing local products	Using environmentally	Using renewable energy	Energy efficiency promotion	Monitoring	Hiring local staff	Staff traning	Wildlife introduction	Education during tours	Waste reducation	Reusing	Recycling	Cultural respect and preservation	Donations	Cleaning up shores	Planting trees	Introducing conservation
1 Abel Tasman Sea Shuttle Limited		✓	✓							✓									
2 Abel Tasman Wilson's Experiences Ltd	✓	✓	✓			✓		✓	✓	✓	✓	✓		✓	✓	✓	✓	✓	
3 Active New Zealand Limited		✓		✓						✓				✓					
4 Ruggedy Range Wildness Experience	✓	✓		✓		✓				✓	✓		✓	✓		✓			
5 Dolphin Blue		✓		✓						✓		✓		✓		✓		✓	
6 Akaroa Dolphins Limited		✓								✓									
7 Auckland Dolphin & Whale Safari (2005) Ltd		✓								✓									
8 Back to Nature Tours New Zealand Limited	✓	✓		✓		✓	✓	✓	✓	✓	✓	✓	✓	✓	✓				✓
9 Beachcomber Cruises(2006) Limited		✓								✓	✓								
10 Black Cat Group 2007 Limited	✓	✓								✓	✓								
11 Dolphin Encounter		✓								✓						✓			
12 Capital Adventure Tours										✓									
13 Carino Sailing & Dolphin Adventures			✓		✓												✓	✓	
14 Seal Swim Kairoura										✓						✓			
15 Collingwood Safari Tours Limited	✓									✓									
16 Crusing Miford Sound Limited		✓								✓									
17 Dive! Tutukaka Limited	✓	✓				✓	✓			✓	✓		✓			✓		✓	✓
18 Dolphin Discoveries(2004) Limited										✓									
19 Dolphin Seafaris(NZ) Limited		✓								✓									
20 Dolphin Watch & Nature Ecotours Ltd	✓	✓							✓	✓									

Contributions / Whale Watching Operators	Eoo-operator	Qualmark	Purchasing and Operation					Staff Management		Education		Waste Control			Conservation Work				
			Purchasing local products	Using environmentally	Using renewable energy	Energy efficiency promotion	Monitoring	Hiring local staff	Staff traning	Wildlife introduction	Education during tours	Waste reducation	Reusing	Recycling	Cultural respect and preservation	Donations	Cleaning up shores	Planting trees	Introducing conservation
21 East Coast Charters										√									
22 Everton Bed and Breakfast		√								√						√			
23 Foxll Sailing Adventures	√									√					√				
24 Kaikoura Kayaks	√	√						√		√									
25 French Pass Motel and Sea Safaris Ltd		√								√									
26 Fullers Bay of Islands Limited	√									√									
27 Wilderness Lodge Lake Moeraki		√								√									
28 Glass Bottom Boat Whitianga Limited		√							√	√									
29 Kaikoura Marine Tours										√									
30 Hollyford Valley Walk Limited		√				√		√		√				√					
31 Aurora Charters										√						√			
32 Onuku Farm Hostel										√									
33 Kahu Kayaks Limited		√								√									
34 Kahurangi Farewell Spit Nature Tours Ltd										√									
35 Kairoura Aero Club Inc										√									
36 Nugget Point and Catlins/Twilight Tours	√									√									
37 Marahau Sea Kayaks										√					√				
38 Abel Tasman Aquataxi	√	√				√	√		√	√	√	√	√	√		√	√		√
39 Marlborough Sounds Adventure Company						√				√		√	√	√		√			
40 Cave Cruzer Adventures	√									√									
41 Moeraki Fishing Charters(2009) Limited										√									
42 Monarch Wildlife Cruises Limited	√									√									

Contributions / Whale Watching Operators	Eoo-operator	Qualmark	Purchasing and Operation					Staff Management		Education		Waste Control			Conservation Work				
			Purchasing local products	Using environmentally	Using renewable energy	Energy efficiency promotion	Monitoring	Hiring local staff	Staff traning	Wildlife introduction	Education during tours	Waste reducation	Reusing	Recycling	Cultural respect and preservation	Donations	Cleaning up shores	Planting trees	Introducing conservation
43 Ngati Koata No Rangitoto Ki Tonga Trust	√									√									
44 Private Discovery Tours	√	√	√	√		√			√	√	√	√		√	√				
45 Kaikoura Pedal Kayaks										√									
46 Pelorous Tours Limited	√									√									
47 White Island Tours Ltd	√	√	√						√	√	√				√	√			
48 Rakiura Charters Limited	√									√	√					√			
49 Real Journeys Limited		√					√			√	√	√		√		√	√		√
50 Seal Coast Safari		√								√									
51 Nine Dives	√									√									
52 Kaikoura Helicopters										√									
53 Kaiteriteris Boat Charter										√									
54 Southern Discoveries				√						√		√	√	√					
55 Topspot Backpackers		√								√									
56 Abel Tasman Charters	√		√	√		√				√	√		√	√	√	√			
57 Catlins Wildlife Trackers	√	√								√	√			√	√				
58 Sails Ashore/ Talisker Charters					√					√									
59 The Sea Kayak Company (2013) limited			√	√	√		√			√				√		√			√
60 Canoes and Kayak Taranaki		√								√									
61 Whale Watch Kaikoura Limited	√	√	√	√			√			√	√	√		√	√	√			√
62 Wild Earth Adventures (2003) Limited										√	√								
63 Wilderness Guides Marlborough Sounds										√									
64 Wings Over Whale Kaikoura Ltd		√								√	√								

5.1 Eco-whale Watching Operators in New Zealand

There were twenty-five operators claiming that they were eco-operators and provided various eco-tours, which accounted for 39.1% of all whale watching operators in New Zealand.

In 1993, ecotourism guidelines for nature tour operators were published by the International Ecotourism Society, establishing a standard for ecotourism (Wood et al., 2002). According to the guidelines, ecotourism operators should take a number of responsibilities. Since the twenty-five whale watching operators label themselves as ecotourism operators, it should mean that they are taking on the responsibilities as ecotourism operators, therefore making contributions to environmental conservation, the development of local communities and the protection of culture.

5.2 Whale Watching Operators with Qualmark Certification

Almost half of the identified operators (thirty-one) exhibit the Qualmark certification on their websites. Qualmark is New Zealand tourism's official quality assurance organisation, and the Qualmark symbol verifies that the business has met stringent quality standards and environmental criteria, thus earning New Zealand tourism's official quality mark (Honey, 2002). Qualmark certified businesses should be able to accomplish a number of minimum requirements, which include various basic environmental practices and completing an environmental review of their business (New Zealand Tourism Guide, 2014).

5.3 Purchasing Local Products

Eight operators (12.5%) stated on their websites that they purchase local goods and services whenever possible. Tourism operations should strive to maximize the economic benefits for the local business and host community, thus a significant proportion of economic advantages should be provided back to local businesses rather than to external companies (Weaver, 2008).

5.4 Using Environmentally Friendly Products

Eleven operators (17.2%) mentioned that they used environmentally friendly products, and most of them have given detailed information with regard to these environmentally friendly products used in their tourism operations. Environmentally friendly products usually mean that they do not harm or damage the environment, whether in their production, use or disposal (Abele et al., 2005).

5.5 Using Renewable Energy

Only three companies noted on their websites that they were utilising renewable energies, which makes only 4.7% of the total sample. These three operators have briefly described the types of renewable energy they were using for their tourism business operations, which are wind energy and solar energy.

According to Fornasiero and Graziani (2012), renewable energy is a clean alternative to fossil fuels, and is better for the environment. Renewable energy technologies are often believed to be eco-technologies due to the fact that they generate clean energy with few (if any) pollutants. The utilization of renewable energy is efficient and aids environmental conservation, which is why it is encouraged in many natural areas for the development of ecotourism (Wood et al., 2002).

5.6 Promoting Energy Efficiency

According to the research findings, ten operators (15.6%) promoted themselves as being energy efficient in their business operations in different ways. As tourism operations and individual visitors make use of energy from either stationary or moving sources, they ultimately contribute to the environmental impact associated with their energy utilization. It could be found that several operators have noticed the importance of developing their vehicles to promote energy efficiency. Some operators use good general operational management to minimize their fuel burn per passenger.

5.7 Monitoring

There were six whale watching operators, which accounted 9.4% of all identified operators, who said that they regularly monitored their performance. It is extremely important for tour operators to minimise their impacts on the environment, hence, the long-term monitoring programs are emphasized as important approaches to assess and minimize human impacts in the principle of ecotourism (Wood et al., 2002). Monitoring can be utilised by tourism operators to assess the scale of both positive and negative effects of their operations on the environment and to predict conditions in the future.

5.8 Hiring Local Staff

Generating employment of local people may be a significant benefit of ecotourism in the local community. The creation of local employment can considerably contribute to that community's social stability and wellbeing (Wood et al., 2002).

As Weaver (2008) noted, local operators should provide various kinds of employment opportunities to local residents. In this study, only four operators (6.3%) mentioned on their websites that they hire local employees.

5.9 Staff Training

Seven operators (10.9%) expressed their views on the importance of staff training. It was found that the training content had focus on two major aspects: environmental knowledge and cultural understandings. The tourism operators are responsible for ensuring that any trip is of a high quality. Giving staff sufficient access to the appropriate training programmes will improve their ability to communicate with and manage tourists in sensitive natural and cultural environments (Wood et al., 2002).

5.10 Wildlife Introduction

All operators provided marine wildlife information since the animals are the primary attractions of their activities. However, some operators provided only little information about the wildlife on their websites, while other operators included detailed information of wildlife. According to a report of IFAW, WWF and WDCS (1997), the marine wildlife that the tourists encounter is the most obvious focus for education, and this education starts with outreach material, such as websites. Through the marine wildlife introductions on the operators' websites, tourists can gain a better understanding about the animals and their habitats, behaviour and characteristics before embarking on the tours.

5.11 Education During Tours

It has been suggested that one of the most significant aspects of whale watching is its potential to educate people, and help them to appreciate and understand marine wildlife and the bio-system they live in (Hoyt, 2007, Lück, 2003). There are nineteen operators, accounting for 29.7% of all companies, claiming that they provide education to tourists during their tours. By studying their websites, it was found that the environment and wildlife conservation are the two main topics of their education programmes. It was noted that some of the operators educate tourists through interpretation. Today, interpretation is a significant educational tool which has been recognised by many whale watching operators for environmental conservation (Markwell & Weiler, 1998).

5.12 Waste Reduction

Twelve operators (18.8%) mentioned that they attempt to reduce the waste they create. As Lück and Higham (2007) indicated, a large amount of marine waste is created by irresponsible whale watching activities and inappropriate tourists' behaviours; this waste has been proven to be detrimental to the marine wildlife and people. According to Weaver (2008), the stakeholders of ecotourism are responsible for their operations to reduce the waste in natural areas, which could have severe impacts on the local ecosystem.

5.13 Reusing

Eight operators, which account for 12.5% of all companies, claimed that they reuse their waste as part of their business operations. Reusing is to use an item again after it has been used. Using products more than once not only reduces the volume of waste, but also the consumption of resources by limiting the demand for high quantities of these products (Beeton, 1998). As Beeton noted, the use of disposable items should be avoided in ecotourism operations. Therefore, reusing products is widely accepted as an economic way in ecotourism businesses.

5.14 Recycling

Sixteen operators asserted to be recycling their waste, which accounts for a relatively high proportion of the sample (25%). By studying the websites with detailed information about recycling, a number of recycled items could be identified. Paper, plastics and glass are the most common items for recycling. There are various benefits of recycling, including decreasing energy usage, reducing the consumption of raw materials and water, minimizing waste, lowering greenhouse gas emissions and even conserving land (Steel, 1999).

5.15 Cultural Respect and Preservation

Nine operators (14.1%) expressed their efforts on the respect and preservation of local culture, and most have clearly indicated how they made contributions to cultural preservation. Local cultural preservation can help a community to not only protect its economically valuable physical assets, but also preserve its practices, history, environment, and a sense of continuity and identity (Wood et al., 2002).

5.16 Donations

Increased financial support is a significant benefit to local environmental conservation, and the various ecotourism stakeholders should make financial contributions through fees, donations, and other forms (Weaver, 2008). Seventeen whale watching operators (26.6%) in New Zealand mentioned that they made financial donations in various forms. A variety of types of trusts, foundations and projects for different conservation purposes are included in their donation contributions. It was found that a majority of these operators chose to give financial support for local wildlife conservation. As Weaver (2008) noted, donating through ecotourism is an effective way to encourage customers to get involved and participate in local conservation efforts.

5.17 Cleaning up Beaches

Only three operators (4.7%) revealed that they regularly help clean up the beaches. Recently, the waste found on beaches has significantly raised public concern about the beach and near-shore water quality. Therefore, effective actions should be taken to resolve this problem by all stakeholders, such as beach clean-ups, an important and useful grassroots initiative arranged and supported by various organisations, including tour operators to conserve the coastal environment (Schwartz, 2005).

5.18 Planting Trees

Based on information collected from the websites, there were only four operators, accounting for 6.3% of all companies, making efforts to plant trees. Locals at tourism destinations and operators can offset tourists' carbon emissions by making financial contributions to projects that address the mitigation of climate impacts. Planting trees is a convenient approach, and one of the most frequently employed methods in carbon offsetting projects (The International Ecotourism Society, 2012).

5.19 Introducing Conservation Programs

According to the research findings, four operators (6.3%) mentioned that they had established their own conservation program. A program was created to offer volunteer working conservation opportunities for tourists. Some operators are working on predator-control programs. Some animals, such as rats, possums and mustelids (ferrets, stoats and weasels), are detrimental to local endangered wildlife (Olstead, et al, 2012). In order to maintain and increase in the populations of protected wildlife, it is paramount to protect them and control their predators.

6. Conclusion

The aim of this study was to provide a comprehensive understanding about the role of whale watching operators in ecotourism in New Zealand through analysing the contents of their websites. By studying the sixty-four operators' websites, a variety of contributions for ecotourism are introduced. Based on the research findings, it appears that some whale watching operators tend to claim on their websites that they are making various efforts to contribute for the development of ecotourism in New Zealand. However, their participation in many of these contributions was found to be relatively low.

This study has only analysed the information about whale watching operators through the data provided on their websites only. It is important to note that this representation on the websites does not necessarily mean that these operators have not more eco-practices in place, and it would be prudent to further study these operators, for example through interviews of managers and staff. Their views concerned with ecotourism and their willingness to make efforts for it are important and should be investigated to gain a greater understanding. Future research is also required to investigate the results or outcomes of these various contributions that operators have made for ecotourism, on the local environment, local communities and tourists' experience, in order to develop their operations for making these measures more efficient and effective.

Acknowledgements

The authors would like to thank David Lundquist at the Department of Conservation in Wellington for his generous support with identifying whale and dolphin watch operators in New Zealand.

References

Abele, E., Anderl, R., Birkhofer, H. (2005). *Environmentally-friendly product development: Methods and tools*. London: Springer.

Babbie, E. (1995). *The Practice of Social Research*. Belmont, CA: Wadsworth.

Beach, D. W., & Weinrich, M. T. (1989). Watching the whales: Is an educational adventure for humans turning out to be another threat for an endangered species? *Oceanus, 2*(32), 84–88.

Beeton, S. (1998). *Ecotourism: A practical guide for rural communities*. Collingwood, Australia: Landlinks Press.

Blamey, R. K. (1995). *The nature of ecotourism.* Canberra: Bureau of Tourism Research.

Blamey, R. K. (1997). Ecotourism: The search for an operational definition. *Journal of Sustainable Tourism,* 5(2), 109–130.

Braun, V., & Clarke, V. (2006) Using thematic analysis in psychology. *Qualitative research in psychology,* 3(2): 77–101

Bryman, A. (2012). *Social research methods.* Oxford: Oxford University Press.

Constantine, R. (1999). *Effects of tourism on marine mammals in New Zealand.* (Vol. 106, Science for Conservation Series). Wellington: Department of Conservation.

Drumm, A., & Moore, A. (2002). *Ecotourism development: A manual for conservation planners and managers.* Arlington: The Nature Conservancy.

Fennell, D. A. (2002). *Ecotourism programme planning.* New York: CBA Publishing.

Fornasiero, P., & Graziani, M. (2012). *Renewable resources and renewable energy: A global challenge.* Boca Raton: CRC Press.

Garin-Munoz, T. & Perez-Amaral, T. (2011). Internet usage for travel and tourism: The case of Spain. Tourism Economics, 17(5): 1071–1085.

Higginbottom, K. (2004). *Wildlife tourism: Impacts, management and planning.* Altona, Vic: Common Ground Publishing.

Higham, J., & Lück, M. (2007). *Marine wildlife and tourism management: Insights from the natural and social sciences.* Wallingford, UK: CAB International.

Honey, M. (1999). *Ecotourism and sustainable development: Who owns paradise?* Washington, D.C.: Island Press.

Hoyt, E. (2007). *A blueprint for dolphin and whale watching development.* Washington, D.C.: Humane Society International (HSI).

Hoyt, E. (2000). *Whalewatching 2000: Worlwide Numbers, Expenditures, and Expanding Socioeconomic Benefits* [webpage]. Retrieved 24.08.2000, from the World Wide Web: http://www.ifaw.org/press/whalewatching2000.html

IFAW (International Fund for Animal Welfare). (1995). *Report of the workshop on the scientific aspects of managing whale watching.* A report for IFAW – the International Fund for Animal Welfare, Dublin, Ireland.

IFAW (International Fund for Animal Welfare), WWF (World Wildlife Fund) & WDCS (Whale and Dolphin Conservation Society) (1997). *Report of the workshop on the educational values of whale watching, Provincetown, Massachusetts, USA*. Retrieved from:

http://www.ifaw.org/sites/default/files/the%20educational%20values%20of%20whale%20watching.pdf

IWC (International Whaling Commission) (1994). Chairman's report of the forty-fifth annual meeting. Appendix 9. IWC Resolution on whale-watching. *Reports of the International Whaling Commission*, 44, 33–34.

Laist, D. W., Knowlton, A. R., Mead, J. G., Collet, A. S., & Podesta, M. (2001). Collisions between ships and whales. *Marine Mammal Science, 17*(1), 35–75.

Lück, M. (2003) Education on marine mammal tours as agent for conservation – but do tourists want to be educated? *Ocean & Coastal Management*, 46(9/10): 943–956.

Lück, M. (2009). *Environmentalism and tourists' experiences on swim-with dolphins tours: A case study of New Zealand*. Saarbrücken, Germany: VDM Verlag.

Lück, M., & Altobelli, R. D. (2009). Coastal tourism as motor for community development: The story of Kaikoura, New Zealand. In R. Dowling & C. Pforr (Eds.), *Coastal tourism development* (pp. 292–306). New York: Cognizant Communication Corporation.

Lück, M., & Higham, J., (2007). Marine wildlife and tourism management: Scientific approaches to sustainable management. In J. Higham & M. Lück (Eds.), *Marine wildlife and tourism management: Insights from the natural and social sciences* (pp. 380–389). Wallingford, UK: CAB International.

Lusseau, D. (2007). Understanding the impacts of noise on marine mammals. In J. Higham & M. Lück (Eds.), *Marine wildlife and tourism management: Insights from the natural and social sciences* (pp. 206–218).Wallingford, UK: CAB International.

Markwell, K., & Weiler, B. (1998). Ecotourism and interpretation. In D. Uzzell & R. Ballantyne (Eds.), *Contemporary issues in heritage & environmental interpretation* (pp. 98–111). London: The Stationary Office.

Mutch, C. (2005). *Doing educational research*. Wellington: NZCER.

New Zealand Tourism Guide. (2014). *Qualmarks*. Retrieved from http://www.tourism.net.nz/qualmark.html

O'Connor, S., Campbell, R., Cortez, H., & Knowles, T., (2009). *Whale Watching Worldwide: Tourism numbers, expenditures and expanding economic benefits*, a special report from the International Fund for Animal Welfare, Yarmouth MA, USA.

Olstead, J., Evans, G., Blades, T., & British Association for Shooting and Conservation. (2012). *Pest & predator control.* Shrewsbury: Quiller.

Orams, M. B. (2000). Tourists getting close to whales, is it what whale-watching is all about? *Tourism Management, 21*(6), 561–569.

Phillips, N. E., & Baird, R. W. (1993). Are killer whales harassed by boats? *The Victorian Naturalist, 50*(3), 10–11.

Schwartz, M. L. (2005). *Encyclopedia of coastal science.* Dordrecht: Springer.

Steel, K. P. (1999). The benefits of recycling. *Science, 285*(5432), 1363–1364.

TIES (The International Ecotourism Society). (2012). *Is carbon offsetting an effective tool for sustainable tourism?* Retrieved 3 December 2012 from https://www.ecotourism.org/news/carbon-offsetting-effective-tool-sustainable-tourism

Vanderlaan, A. S. M., & Taggart, C. T. (2007). Vessel collisions with whales: The probability of lethal injury based on vessel speed. *Marine Mammal Science, 23*(1), 144–156.

Weaver, D. B. (1998). *Ecotourism in the less developed world.* Wallingford: CABI Publishing.

Weaver, D. B. (2008). *Ecotourism.* Milton, Qld: John Wiley & Sons Australia.

Weber, R. (1990). *Basic content analysis* (2nd ed.). Newbury Park, CA: Sage.

Western, D. (1993). Defining ecotourism. In K. Lindberg & D. E. Hawkins (Eds.), *Ecotourism: A guide for planners and managers* (pp. 7–11). North Bennington, Vermont: The Ecotourism Society.

Wood, M. E., United Nations Environment Programme & International Ecotourism Society. (2002). *Ecotourism: Principles, practices & policies for sustainability.* Burlington, VT: UNEP Division of Technology, Industry and Economics.

Woods-Ballard, A. J., Parsons, E. C. M., Hughes, A. J., Velander, K. A., Ladle, R. J., & Warburton, C. A. (2003). The sustainability of whale-watching in Scotland. *Journal of Sustainable Tourism, 11*(1), 40–55.

Pascal Languillon

Green Motivation and Luxury Travel

1. Introduction

We live at a time when environmental awareness is growing among many people. As a result, companies are developing green products and are attempting to stimulate demand through green marketing. Luxury hotels, which are often synonymous with conspicuous consumption, are also embracing the 'green wave' and developing comprehensive environmental and social policies (ESP). Despite decades of academic research in the fields of green consumption, tourism motivation and ecotourism, little is known about the influence of green factors on tourists' buying decision processes. This paper looks at the literature and demonstrates why further research is needed in this area.

2. The Emergence of a New Luxury

The word luxury comes from the Latin 'luxuria', which means 'excess' or 'extras of life' Some dictionaries define luxury as "something that is considered an indulgence rather than a necessity" (Collins, 2002) whereas others stress out that a luxury is "something expensive or hard to obtain" (The American Heritage, 2000). While it is always linked to the generation of pleasure, satisfaction and comfort, luxury can also be viewed as "a foolish or worthless form of self-indulgence" (Dictionary.com Unabridged, 2010). Danziger (2005, p. 17) notes that "widely recognized as the opposite of necessity, luxury is an emotionally charged word that carries a lot of baggage along with it."

As early as 1899, the American economist Thorstein Veblen introduced the term 'conspicuous consumption' in his work *The Theory of the Leisure Class*. Noting the emergence of an unproductive leisure class that accumulates wealth not enjoyed by the labouring class, he argued that members of this leisure class convey their position of wealth and status to other members of society through conspicuous consumption. Veblen contended that luxury not only fulfils hedonistic expectations, it also – and mostly – enables people to display income and social power. He criticized the consumption of luxury items for being a narcissistic behaviour. At a time of global ecological crisis, when people are encouraged to "tread lightly on Earth", luxury is also criticized for its high environmental impacts (Brace, 2007). Luxury is usually associated with abundance and space – a contradiction in a world of depleting resources and increased population density. Dubois and Ceron (2008) argue that people flying in business or first-class have a higher carbon footprint than passengers in economy seats by using more room in the planes, thus preventing airlines to reduce their number of flights by carrying more passengers per flight. Gossling (2002) and Brenner and Aguilar (2002) point out that luxury resorts in developing nations create enclave-like pockets of wealth that do not benefit local communities and put a heavy burden on local resources. Bloggers recently mocked climate change campaigner and former US Vice-President Al Gore for the recent acquisition of a huge mansion in California – deliberately increasing his personal carbon footprint for the sake of living in a lavish environment (Hyperbole Media, 2010).

Conspicuous consumption is not 'hip' anymore, and this became especially relevant in the wake of the current global economic crisis, as noted by Danziger (2007, p. 43): "Luxury is no longer about an external show of status or wealth, but an inner state of being defined by personal happiness and an outstanding quality of life." The luxury consumer is indeed changing, and 'new luxury' is more about the experience than the traditional status and prestige ideal of 'old luxury' (Silverstein, Fiske, & Butman, 2004). Danziger (2005) studied the preferences of more than a thousand American luxury consumers and concluded that they are more interested in living outstanding experiences such as luxury travel than about acquiring more 'stuff'. She also reported a fundamental cultural shift:

> These changes in how people look at luxury and their new value system that rejects rampant materialism will be lasting. The new generation of consumers are thinking about the future in new ways. They are learning that conspicuous consumption is not the way to grow wealth, to achieve happiness or to make their life more meaningful. Affluents who have much are giving back, doing good and taking concrete steps to making the world a better place. (p. 25)

Affluent consumers are becoming concerned about the way companies do business and deliver their services. They expect the companies they support to be good stewards of the environment by implementing green practices and to be good citizens of the world by giving back and donating funds to charities (Danziger, 2005; Honey & Gilpin, 2009). People aspire to imitate the elites, and Hollywood stars have certainly been instrumental in fuelling the growing green demand. Some websites have specialized in writing stories about how the rich and famous green their lifestyle (Planet Green, 2010). Consumer demand for sustainable products is growing, and companies have a responsibility to meet that demand, as noted in a study commissioned by the World Wildlife Fund to see how luxury brands integrate sustainability in their businesses (Bendell & Kleanthous, 2007, p. 2): "Luxury brands have the power to influence consumer aspiration and behaviour by editing consumer choices through product design, distribution and marketing; and by influencing how, when and for how long consumers use their products. They have both the opportunity and the responsibility to promote sustainable consumption."

As the concept of luxury means different things to different people, the perception of what constitutes a 'luxury hotel' is also highly subjective. Hotel managers often turn this to their advantage as many relatively modest hotels and B&Bs market themselves as being 'luxury'. According to industry specialists, the word 'luxury' is one of the most abused words in the hospitality lexicon (Meetings Net, 2006).

Luxury hotels are considered to be the finest hotels in the world, providing the highest standards of comfort and service, and typically rated as being 'five-star' or above by national quality accreditation schemes. These properties typically maintain a high staff-to-guest ratio, gourmet dining, and 24-hour room service. This quest for excellence comes at a price – for the clients as well as for the environment: "With a high daily turnover of guests paying for the highest standards, multiple choices and generous allocations in personalised services, room furnishings, laundry, food and appliance usage, luxury hotel operations use significant amounts of water and energy, and produce large amounts of waste." (Brace, 2007, p. 136). Many share the view that "luxury hotels are the great sinners when it comes to the environment" (Mc Cracken cited in Brace, 2007, p. 136). As more and more luxury hotels embrace sustainability in their business practices, this view becomes challenged. A healthy social and natural environment is the first and most important resource of the tourism industry. It makes long term business sense for tourism stakeholders to 'tread lightly' on the planet, and as consumer demand is growing, many hotels no longer contend with simple measures such as giving guests the option of not having their towels changed daily but now seek to create a holistic 'green experience', as noted by Languillon, O'Brien and Zee (2009):

> Ecochic luxury, where travellers enjoy small-scale properties located in unspoiled environments and experience the finest architecture and service indigenous and local communities have to offer, is here. Green city hotels, safari camps and ecolodges are booming all around the world, offering a holistic experience for travellers in search of relaxation, fulfilment, discovery, adventure and most of all a deep connection to the planet, and to its people. (p. 24)

As the majority of the world's population is now urbanized and surrounded by concrete walls, access to wilderness – once disrespected and feared – becomes a real luxury. People who live in cities now seek refuge from modernity in the natural environment; they have developed a romance for nature. The well-to-do are no different and wish to be luxuriously at one with nature, immersed in peaceful and quiet surroundings. In the words of Marriott International President Arne Sorenson, "conspicuous consumption and lavish luxury hotels are no longer in vogue with travellers" (Seetharaman, 2010).

Some hotel companies have also gone to great lengths to improve the quality of life in local communities and the well-being of their employees, for instance dedicating a percentage of the profits to local charities and ensuring educational opportunities for their staff (Bohdanowicz & Zientara, 2009). Caring for the environment and the local communities is not just good for the planet, it also makes business sense as noted by Oines and Assenov (2005, p. 645): "eco-friendly practices will offset initial costs due to increased efficiency, additional guests and the possibility to charge a premium". A recent study showed that socially responsible activities helped U.S. hotels achieve their financial goals (Lee & Park, 2009). This is confirmed by findings that companies focused on sustainability outperformed their peers by 15% during the first year of the financial crisis (A.T. Kearney, 2009).

There are therefore many reasons why luxury hotels should embrace sustainability, and indeed many have gone that route in recent years, with initiatives to reduce energy consumption and carbon emissions, minimize waste and water usage, and conserve the surrounding ecosystems (Bohdanowicz, 2006; Erdogan & Baris, 2007; Mensah, 2006; Ryan & Stewart, 2009). However despite hotels' rising financial investments in designing sustainable policies and green marketing, little is known about the impacts of these efforts on sales.

3. Profile of the Green Traveller

With over five thousands articles to date in the academic literature, the interconnected notions of 'ecotourism', 'sustainable tourism' and 'green travel' have become some of the most popular topics of enquiry among tourism studies (Baretje-Keller, personal communication, 5 Jan 2011). According to The International Ecotourism Society (1992), ecotourism is "responsible travel to natural areas that conserves the environment and improves the well-being of local people". Sustainable tourism and green tourism are terms used more broadly to refer to tourism that has positive impacts (or that at least mitigates negative impacts) on the economy, the local community and the environment, but that does not necessarily take place in the natural environment (i.e. city hotels can also be sustainable if they minimize their negative impacts).

Travellers believed to partake in sustainable tourism activities have been referred to as 'ecotourists', 'green tourists', 'responsible travellers', 'ethical travellers' or 'environmentally-friendly tourists'. A number of studies have attempted to profile ecotourists and green travellers (CMIGreen, 2009; Eagles & Cascagnette, 1995; Han, Hsu, Lee, & Sheu, 2010; Kerstetter, Hou, & Lin, 2004; Wight, 2001), and the focus has been almost entirely on market segmentation, based on geographic, demographic and behavioural characteristics. Epler Wood (2004, p. 2) simply defined ecotourists as "people who travel with ecotourism as their primary motivation", while Eagles and Cascagnette (1995, p. 22) suggested that ecotourists are individuals who "travel with the intent of observing, experiencing and learning about nature". Ecotourists are generally believed to be middle-aged, to have higher levels of education and income than average travellers and to mostly originate from developed countries (Ballantine & Eagles, 1994; Chafe, 2007; Epler Wood, 2004). Authors have differentiated ecotourists from a behavioural perspective between hard and soft ecotourists (Diamantis, 1999; Orams, 2001), passive and active ecotourists (Orams, 2001; Wight, 2001), and casual and frequent ecotourists (Diamantis, 1999).

Various consumer surveys have attempted to provide an understanding of the size of the ecotourist market (Butcher, 2006; CMIGreen, 2009; Wight, 1996). The United Nations World Tourism Organization has stated for years that ecotourism was the fastest growing area of tourism (Kester, 2002), information that has been reproduced extensively in numerous articles even though it is now believed to be a myth: Epler Wood (2004, p. 2) noted that "ecotourism demand studies have been very optimistic over the years", while Weaver and Lawton (2007, p. 1171) contended that "attempts to isolate ecotourists as a distinct set of consumers are constrained by disagreements over the parameters of ecotourism itself".

It is also difficult to define ecotourists because their motivations overlap with those of other types of tourists. Wight (2001) stated that ecotourism markets are not homogenous, and more than ten years on, it seems like there is still no consensus over the size of these markets. Some authors now argue that ecotourists should not even be considered to be a specific market. A recent survey on visitors to a National Park in Australia showed that less than six per cent of visitors could be considered 'true ecotourists', concerned whether their experience conforms to the three core criteria of ecotourism (Beaumont, 2010). Sharpley (2006, p. 19) went further and advocated the need to stop referring to the term 'ecotourist':

> The nature of tourism consumption is such that environmental concern is likely to be subordinated to a variety of other values, motivations and desired benefits – in other

> words, tourists may be demanding new, different products, such as ecotourism, yet the conventional reasons for participating in tourism are unlikely to have changed. In short, there is little distinction between the 'ecotourist' and the 'mass' tourist. Thus, it is, perhaps, inappropriate to continue to refer to the 'ecotourist'; it is a label that is becoming increasingly meaningless.

Recent studies have tried to assess who the 'green tourist' is (Andereck, 2009; Bergin-Seers & Mair, 2009; CMIGreen, 2009; Yabuta & Scott, 2010). Bergin-Seers and Mair (2009, p. 117) contend that "green tourists exist, even though they may be inconsistent in their behaviour". The authors view green tourists as

> tourists who are influenced by their environmental beliefs and practices; who are interested in being environmentally friendly on holiday; at times select holidays by considering environmental issues; and are potentially willing to pay extra for products and services provided by environmentally friendly tourism operators (p. 117).

People with a pro-environmental attitude may choose to patronize a green hotel one day but stay in a conventional hotel the following day: travellers do not belong to artificial typologies and their attitudes and behaviours vary greatly depending on situational variables. Can people who fly to the other side of the world to partake in ecotourism activities really be called 'green travellers'? As transport to the destination often constitutes the biggest part of the environmental footprint of tourists (Gossling & Peeters, 2007; Peeters & Schouten, 2006), some authors contend that the greenest travellers would actually be those who stay at home (Bergin-Seers & Mair, 2009).

After reviewing 19 journal articles about eco-friendly travellers from top academic journals, Dolnicar et al. (2008) recommended destination managers to target "tourists with low ecological footprints", those who stated that maintaining an unspoiled environment on vacation is important to them. This view is in itself very questionable and symptomatic of the current confusion surrounding the issue of green travellers in the academic field. Tourists who care about the environment may in many cases have bigger ecological footprints than those who do not, because they go further away from big cities in order to get the natural experience they are looking for, generating more important carbon emissions and leaving impacts in previously untouched or fragile environments. For instance cultural tourists interested in visiting cities and museums certainly have a smaller ecological footprint than birdwatchers that travels the globe in order to sight rare species. If it is true that an ecotourism supply exists (e.g. ecolodges), the term 'ecotourists' has to be used with caution. In all fairness, there are no totally green travellers, only people who take into account eco-social standards when selecting their travel provider. Identifying and targeting tourists who have a disposition to buy green products and book green hotels can lead to increased sales – thus further motivating hotels to adopt sustainable practices.

4. Green Motivation

While the concepts of 'green traveller' and 'ecotourist' have to be used circumspectly, and the results of market segmentation studies taken with caution, the fact that a considerable number of people are concerned about sustainability within the context of travel and tourism cannot be disputed (CMIGreen, 2009; Mehmetoglu, 2010). Even though people may not choose their hotels primarily based on sustainability criteria, more and more travellers express concerns about green aspects of their trips and deliberately choose to book 'eco-lodges' and other green hotels (CMIGreen, 2009). Sustainability can be in some cases part of the mental equation that leads to the decision of booking a hotel room. There is also evidence that some tourists may choose an airline based on its environmental management practices (Horner and Swarbrooke, 2006). It is important to bear in mind that green tourists are not a homogenous group. Authors have used the term 'shades of green' to describe the different types of green consumers – from 'not green at all' to 'dark green' (CMIGreen, 2009; Horner & Swarbrooke, 2007). Even 'the dark green consumer' only considers green factors in a sporadic fashion, depending on the circumstances (Young, Hwang, McDonald, & Oates, 2010).

When the term 'green consumers' started to emerge at the end of the 1980s, it was to describe people who avoid buying products that have negative impacts (Elkington & Hailes, 1989). Among the main concerns at the time were animal cruelty, rainforest protection and ozone depletion; enabling cosmetics companies like 'The Body Shop' to thrive because they were 'cruelty-free'. The main environmental concerns have evolved and have since incorporated issues such as climate change and organic food. Green tourists have become more informed over the years and are now concerned with a wide array of issues regarding their trip and accommodation, including issues of resource use, pollution, recycling, conservation and energy (Horner & Swarbrooke, 2007). There have been attempts in the literature to segment consumers according to the extent of their environmental concerns, but there have been many conflicting results, as beliefs do not necessarily translate into purchasing (Peattie, 2001). This wide gap observed between pro-environmental attitude and pro-environmental action has been very well documented (Boulstridge & Carrigan, 1993; Epler Wood, 2004; Lane & Potter, 2007; Vermeir & Verbeke, 2006). Green consumption is indeed a complex subject of enquiry, as noted by Bergin-Seers and Mair (2009, p. 110): "A problem with much of the research about green consumerism is the complexity surrounding consumer behaviour; the enormity of the range of influences on consumers and the way these affect individuals over time". Among those influences are attitudes, social norms, and the perceived difficulty of performing green behaviour (Han, Hsu, Lee, & Sheu, 2010).

Considerable attention has been given to the issue of green consumption, but relatively few studies have attempted to understand the underlying reasons of green behaviour, i.e. why people choose to buy green products (Peattie, 2010). It is believed that there is a mix of rational and emotional reasons for 'going green'. Ottman (1992) described green consumers as individuals who are mainly looking to protect themselves and their world through the power of their purchases. Organic food is for instance mainly bought to enhance or preserve health (Hammitt, 1990). Horner and Swarbrooke (2007, p. 181) note that "green tourists might be influenced by a number of motivators including an altruistic belief in the need to protect the environment, a desire to feel good about their behaviour as tourists, and a wish to improve their image among friends and relatives." Green tourists can also desire to have positive impacts on the local economy and help alleviate poverty in host communities (Chafe, 2007).

Many authors share the view that the main causes of green behaviour are not altruistic (Horner & Swarbrooke, 2007; Young, et al., 2010). Cater (2006) contends that ecotourists visit exclusive lodges and far flung destinations in order to gain social capital. There are reasons to believe that people mainly go green to improve their self-image (the global perception of themselves and what they would like to be) and their social self-image (how they are and would like to be perceived) (Font & Buckley, 2001). This would tend to reflect a need for achievement, power and status, contrary to Doran's (2009) belief that universalism values – defined by Schwartz (1994, p. 12) as "understanding, appreciation, tolerance and protection, for the welfare of all people and for nature" – are the best predictors of eco-social consumption.

5. Tourism Motivation

There is a need to understand tourists and their purchasing behaviour in order to properly plan promotional activities, forecast market trends, develop new products and meet economic development goals. As noted by Cohen (2006, p. 1), "to successfully compete for customers' wallet share, you must connect with them, understand their buying habits, and anticipate their every move." The understanding of customer needs and wants should come before any other marketing activity, but in reality, many companies have learnt to convince consumers by trial and error rather than having a thorough understanding of the complex purchasing processes (Decrop, 2006; Laws, 2002). Academics have produced various tourism motivation models that show that tourists follow a decision making process that involves the following steps: need recognition, search for information, evaluation of alternatives, purchase, and feedback.

Schmoll (1977) created one of the first tourism motivation models and showed that traveller decision making is influenced by travel stimuli (e.g., advertising), personal and social determinants (e.g., socio economic status) and external variables (e.g., constraints of time). Moutinho (1986) perhaps provided the most complex model to date and emphasized that travel decisions are heavily affected by social influences such as family and culture. Woodside and MacDonald (1994) were the first researchers to recognize that destination choice is one of many travel related decisions one has to make (e.g. activities, travel routes, accommodation, dining), and that choices amongst these vacation elements are largely interdependent. All these theories have started to shape an understanding of tourism motivation, but Horner and Swarbrooke (2007) assert that the best known models are incomplete, and that the fact that they are at least twenty years old is a significant weakness in such a dynamic industry. Sirayaka and Woodside (2005, p. 816) reviewed the different models and concluded that it is unlikely that "individual decision processes fit neatly into a single decision theory". Most models see the decision as a linear process and no distinctions are made about which factors might weight heavier than others when decisions are being made. None of the current tourism motivation models consider sustainability as a potential decision factor.

Factors which influence the purchase of a particular tourism experience can be divided into determinants and motivators (Horner & Swarbrooke, 2007). The determinants are the factors which determine to what extent tourists are able to purchase the service they desire. They can be constraints or facilitators and either personal (disposable income, time, past travel experience) or external (views of friends and relatives, influences of the media). The motivators are the factors which motivate tourists to choose a destination and to buy tourism products. The distinction has been drawn between internal 'push factors', which push people to get away and motivate them to take a holiday (e.g., desire for relaxation, for romance) and external 'pull factors', which attract people towards a particular destination or product at a particular time (e.g., cultural and natural attractions) (Crompton, 1979). Tourists are influenced by a number of motivators at any one time, and their holidays represent a compromise between their multiple motivators. Price, the quality of facilities and the quality of the destination's environment are usually prime motivators, but some travellers report that their hotels' environmental and social policies are part of their mix of motivators (CMIGreen, 2009; Goodwin & Francis, 2003).

Correia and Moital (2009, p. 24) note that "the vast majority of tourism motivation studies have focused on a comprehensive understanding of what motivates the individual as a whole [and that] the in-depth study of the individual motivations of tourists is rare". Only few studies have loosely assessed how green factors fit into tourism motivation theories and models (Luo & Deng, 2008).

6. Conclusion

Despite decades of research on ecotourism (Fennell & Eagles, 1990; Honey, 1999; Weaver & Lawton, 2007), there is still little insight into who eco-friendly tourists really are (Dolnicar, et al., 2008). Extensive research work on tourist motivators has been documented in the literature (Dann, 1981; Gnoth, 1997; Pearce & Lee, 2005), but there is a lack of studies on the importance of sustainability as a motivation factor (Holden & Sparrowhawk, 2002; Luo & Deng, 2008). Despite growing evidence that ethical factors play an increasing role in the vacation decision making process (CMIGreen, 2009), a thorough understanding of green motivation as a purchase factor is still missing (Budeanu, 2007; Han, Hsu, & Sheu, 2010), and this is particularly true in the luxury travel sector. Further research is therefore needed.

References

A.T. Kearney. (2009). *Green Winners, the performance of sustainability-focused companies during the financial crisis*. Retrieved 22 April 2010, 2010, from http://www.atkearney.com/images/global/pdf/Green_winners.pdf.

Andereck, K. L. (2009). Tourists' perceptions of environmentally responsible innovations at tourism businesses. *Journal of Sustainable Tourism, 17*(4), 489–499. doi:10.1080/09669580802495790

Ballantine, J. L., & Eagles, P. F. J. (1994). Defining canadian ecotourists. *Journal of Sustainable Tourism, 2*(4), 210–214.

Beaumont, N. (2010). Identifying the ecotourist market using the core criteria of ecotourism: is there a true ecotourist that seeks nature, learning and sustainability?*Council for Australian Tourism and Hospitality Education (CAUTHE)*.

Bendell, J., & Kleanthous, A., &. (2007). *Deeper Luxury: Quality and style when the world matters*: WWF-UK http://www. wwf. org. uk/deeperluxury/report.html.

Bergin-Seers, S., & Mair, J. (2009). Emerging green tourists in Australia: Their behaviours and attitudes. *Tourism and Hospitality Research, 9*(2), 109–119.

Bohdanowicz, P. (2006). Environmental awareness and initiatives in the Swedish and Polish hotel industries—survey results. *International Journal of Hospitality Management, 25*(4), 662–682.

Bohdanowicz, P., & Zientara, P. (2009). Hotel companies' contribution to improving the quality of life of local communities and the well-being of their employees. *Tourism and Hospitality Research, 9*(2), 147–158.

Boulstridge, E., & Carrigan, M. (1993). Do consumers really care about corporate responsibility? Highlighting the attitude—behaviour gap. *Journal of Communication Management, 4*(4), 355–368.

Brace, M. (2007). The lapse of luxury. *ECOS, 2007*(136), 14–17.

Brenner, L., & Aguilar, A. G. (2002). Luxury tourism and regional economic development in Mexico. *The Professional Geographer, 54*(4), 500–520.

Budeanu, A. (2007). Sustainable tourist behaviour – a discussion of opportunities for change. *International Journal of Consumer Studies, 31*(5), 499–508. doi:10.1111/j.1470-6431.2007.00606.x

Butcher, J. (2006). The United Nations International Year of Ecotourism: a critical analysis of development implications. *Progress in Development Studies,* 6(2), 146.

Cater, E. (2006). Ecotourism as a Western construct. *Journal of Ecotourism,* 5(1), 23–39.

Chafe, Z. (2007). Consumer demand for quality in ecotourism. *Quality Assurance and certifications in Ecotourism*, 164–195.

CMIGreen. (2009). *Green Traveler Survey Report.* San Francisco, CA.

Cohen, M. (2006). *Why customers do what they do: Who they are, why they buy, and how you can anticipate their every move*: McGraw-Hill Companies.

Collins. (Ed.) (2010). Retrieved on 20 October 2010 from the website http://www.collinsdictionary.com/dictionary/english/luxury.

Correia, A., & Moital, M. (2009). The Antecedents and Consequences of Prestige Motivation in Tourism: An Expectancy-Value Motivation.

Crompton, J. L. (1979). Motivations for pleasure vacation. *Annals of Tourism Research,* 6(4), 408–424.

Dann, G. (1981). Tourist motivation an appraisal. *Annals of Tourism Research,* 8(2), 187–219.

Danziger, P. (2005). *Let them eat cake: marketing luxury to the masses-as well as the classes.* Chicago, IL: Kaplan Publishing.

Danziger, P. (2007). New Luxury. *Global Cosmetic Industry, 175*(5), 42.

Decrop, A. (2006). *Vacation decision making.* Wallingford, Oxford, UK: CABI.

Diamantis, D. (1999). The characteristics of UK's ecotourists. *Tourism Recreation Research, 24*(2), 99–102.

Dictionary.com Unabridged. (2010). *luxury. (n.d.).* Retrieved October 21, 2010, from http://dictionary.reference.com/browse/luxury.

Dolnicar, S., Crouch, G. I., & Long, P. (2008). Environment-friendly Tourists: What Do We Really Know About Them? *Journal of Sustainable Tourism, 16*(2), 197–210. doi:10.2167/jost738.0

Doran, C. J. (2009). The role of personal values in fair trade consumption. *Journal of business ethics, 84*(4), 549–563.

Dubois, G., & Ceron, J. P. (2008). Carbon labelling and restructuring travel systems: Involving travel agencies in climate change mitigation. *Sustainable Tourism Futures: Perspectives on Innovation, Scale and Restructuring from Gössling, S., Hall, M. & Weaver, D.(eds.), Advances in Tourism Research. Elsevier.*

Eagles, P. F. J., & Cascagnette, J. W. (1995). Canadian ecotourists: who are they? *Tourism Recreation Research, 20*(1), 22–28.

Elkington, J., & Hailes, J. (1989). *The green consumer guide*: Penguin.

Epler Wood, M., &. (2004). *The Green Market Gap, Epler Wood Reports. Epler-Wood International, Burlington, Vermont. Taken from results of International Market Study for Awa/Huaorani Territories, Ecuador, Proyecto Caiman, US-AID, Quito, Ecuador.*

Erdogan, N., & Baris, E. (2007). Environmental protection programs and conservation practices of hotels in Ankara, Turkey. *Tourism Management, 28*(2), 604–614.

Fennell, D. A., & Eagles, P. F. J. (1990). Ecotourism in Costa Rica: A conceptual framework. *Journal of Park and Recreation Administration, 8*(1), 23–34.

Font, X., & Buckley, R. (2001). *Tourism ecolabelling: certification and promotion of sustainable management*: CABI.

Gnoth, J. (1997). Tourism motivation and expectation formation. *Annals of Tourism Research, 24*(2), 283–304.

Goodwin, H., & Francis, J. (2003). Ethical and responsible tourism: Consumer trends in the UK. *Journal of Vacation Marketing, 9*(3), 271.

Gossling, S. (2002). Global environmental consequences of tourism. *Global Environmental Change, 12*(4), 283–302.

Gossling, S., & Peeters, P. (2007). It Does Not Harm the Environment! An Analysis of Industry Discourses on Tourism, Air Travel and the Environment. *Journal of Sustainable Tourism, 15*(4), 402–417. doi:10.2167/jost672.0

Hammitt, J. K. (1990). Risk Perceptions and Food Choice: An Exploratory Analysis of Organic Versus Conventional Produce Buyers. *Risk Analysis, 10*(3), 367–374.

Han, H., Hsu, L. T. J., Lee, J. S., & Sheu, C. (2010). Are lodging customers ready to go green? An examination of attitudes, demographics, and eco-friendly intentions. *International Journal of Hospitality Management.*

Han, H., Hsu, L. T. J., & Sheu, C. (2010). Application of the theory of planned behavior to green hotel choice: Testing the effect of environmental friendly activities. *Tourism Management, 31*(3), 325–334.

Holden, A., & Sparrowhawk, J. (2002). Understanding the motivations of ecotourists: the case of trekkers in Annapurna, Nepal. *International Journal of Tourism Research, 4*(6), 435–446.

Honey, M. (1999). *Ecotourism and sustainable development: who owns paradise?*: Island Pr.

Honey, M., & Gilpin, R. (2009). Tourism in the Developing World.

Horner, S., & Swarbrooke, J. (2007). *Consumer Behaviour in Tourism.*

Hyperbole Media. (2010). *Al Gore increases his carbon footprint, buys house in ritzy Santa Barbara neighborhood.* Retrieved 21 October 2010, 2010, from http://www.ihatethemedia.com/al-gore-buy-home-in-montecito.

Kester, J. G. C. (2002). WTO Tourism Market Trends, Edition 2002 and the International Year of Ecotourism 2002. *Tourism Economics, 8*(4), 463–477.

Lane, B., & Potter, S. (2007). The adoption of cleaner vehicles in the UK: exploring the consumer attitude-action gap. *Journal of Cleaner Production, 15*(11–12), 1085–1092.

Languillon, P., Zei, T. M., & Kate, O. B. (2009). *Ecochic* (Editions Didier Millet ed.). Singapour

Laws, E. (2002). *Tourism marketing: quality and service management perspectives*: Continuum International Publishing Group.

Lee, S., & Park, S. Y. (2009). Do socially responsible activities help hotels and casinos achieve their financial goals? *International Journal of Hospitality Management, 28*(1), 105–112.

Luo, Y., & Deng, J. (2008). The new environmental paradigm and nature-based tourism motivation. *Journal of Travel Research, 46*(4), 392.

Meetings Net. (2006, Dec 18, 2006). *Redefining the High End: Ritz-Carlton's Luxury Meetings Forum*. Retrieved 18 January 2011, from http://meetingsnet.com/news/ritz_carlton_meetings_forum/.

Mehmetoglu, M. (2010). Accurately Identifying and Comparing Sustainable Tourists, Nature-Based Tourists, and Ecotourists on the Basis of Their Environmental Concerns. *International Journal of Hospitality & Tourism Administration, 11*(2), 171–199. doi: 10.1080/15256481003732840

Mensah, I. (2006). Environmental management practices among hotels in the greater Accra region. *International Journal of Hospitality Management, 25*(3), 414–431.

Moutinho, L. (1986). *Consumer behaviour in tourism*: MCB University Press. doi:10.1108/EUM0000000004718

Oines, A., & Assenov, I. (2005). *Competite advantage through developing environmentally friendly hotel resorts*. Forum proceedings, The Fifth Asia Pacific Forum for Graduate Student Research in Tourism. Faculty of Service Industries, Prince of Songkla University. Phuket.

Orams, M. B. (2001). Types of ecotourism. *The encyclopedia of ecotourism*, 23–36.

Ottman, J. (1992). Sometimes consumers will pay more to go green. *Marketing News, 6*.

Pearce, P. L., & Lee, U.-I. (2005). Developing the Travel Career Approach to Tourist Motivation. *Journal of Travel Research, 43*(3), 226–237. doi:10.1177/0047287504272020

Peattie, K. (2001). Golden goose or wild goose? The hunt for the green consumer. *Business Strategy and the Environment, 10*(4), 187–199.

Peattie, K. (2010). Green Consumption: Behavior and Norms. *Annual Review of Environment and Resources, 35*(1), 195–228. doi:doi:10.1146/annurev-environ-032609-094328

Peeters, P., & Schouten, F. (2006). Reducing the Ecological Footprint of Inbound Tourism and Transport to Amsterdam. *Journal of Sustainable Tourism, 14*(2), 157–171.

Planet Green. (2010). *Hollywood Green*. Retrieved 18 January 2011, from http://planetgreen.discovery.com/tv/hollywood-green/.

Ryan, C., & Stewart, M. (2009). Eco-tourism and luxury–the case of Al Maha, Dubai. *Journal of Sustainable Tourism, 17*(3), 287–301. doi:10.1080/09669580802366587

Schmoll, G. A. (1977). *Tourism promotion*: Tourism international.

Schwartz, S. H. (1994). Are there universal aspects in the structure and contents of human values? *Journal of social issues, 50*(4), 19–45.

Seetharaman, D. (2010). *Lavish hotels out of vogue: Marriott*. Retrieved 18 January 2011, from http://www.reuters.com/article/idUSTRE61N3QL20100225.

Sharpley, R. (2006). Ecotourism: A Consumption Perspective. *Journal of Ecotourism, 5*(1), 7–22. doi:10.1080/14724040608668444

Silverstein, M. J., Fiske, N., & Butman, J. (2004). Trading up: why consumers want new luxury goods-and how companies create them.

Sirakaya, E., & Woodside, A. G. (2005). Building and testing theories of decision making by travellers. *Tourism Management, 26*(6), 815–832. doi:doi:10.1016/j.tourman.2004.05.004

The American Heritage. (2010). Retrieved on 20 October 2010 from the website ahdictionary.com.

The International Ecotourism Society (2010). Retrieved October 21, 2010 from the website http://www.ecotourism.org.

Veblen, T. (1899). *The theory of the leisure class*: Oxford University Press, USA.

Vermeir, I., & Verbeke, W. (2006). Sustainable food consumption: exploring the consumer "attitude–behavioral intention" gap. *Journal of Agricultural and Environmental Ethics, 19*(2), 169–194.

Weaver, D. B., & Lawton, L. J. (2007). Twenty years on: The state of contemporary ecotourism research. *Tourism Management, 28*(5), 1168–1179. doi:doi:10.1016/j.tourman.2007.03.004

Wight, P. A. (1996). North American ecotourists: Market profile and trip characteristics. *Journal of Travel Research, 34*(4), 2.

Wight, P. A. (2001). Ecotourists: Not a homogeneous market segment. *The encyclopedia of ecotourism*, 37–62.

Woodside, A. G., & MacDonald, R. (1994). General system framework of customer choice processes of tourism services. *Spoilt for choice*, 30–59.

Yabuta, M., & Scott, N. (2010). Ecotourist Behaviour as a Green Consumer.

Young, W., Hwang, K., McDonald, S., & Oates, C. J. (2010). Sustainable consumption: green consumer behaviour when purchasing products. *Sustainable Development, 18*(1), 20–31. doi:10.1002/sd.394

Destination Branding and Economic Impacts

Anne Köchling, Bernd Eisenstein and Alexander Koch

On Theme Marketing of Tourism Organisations: "Destination Brand" – Analysis of the Theme Competence of German Holiday Destinations

1. Introduction

As part of strategic marketing, the majority of German tourism organisations focus on touristic themes or types of holiday (theme marketing). Against this background, an independent study was carried out to determine theme competence from the point of view of the German population. This study, "Destination Brand – Theme competence of German holiday destinations", ascertained theme competence for holiday destinations in German domestic tourism on an almost nationwide scale for the first time.[1] It was carried out primarily in October 2010 and repeated three years later with the same methodology in order to obtain a time comparison.

The results of the study are representative for the German-speaking population aged between 14 and 74.[2]

In the study, the general interest of the respondents in different holiday themes or holiday activities was surveyed. Furthermore, for each of the destinations considered, the respective eligibility of selected holiday themes was analysed. As data was collected for a total of 122 German destinations in the first study in 2010 and for even more destinations in 2013, the study also provided the chance to make a detailed comparison of the competitors (benchmarking). The data gathered in the course of the study, allows a data based transfer of combinations of target groups and the range of services in specific destinations and hereby provides a sound basis for strategic decisions with regard to implementing the target group-oriented theme marketing of the destination.

---

1 The study was carried out by the Institute of Management and Tourism (IMT) of the Westcoast University of Applied Sciences in cooperation with the market research company GfK-Travelscope in Nürnberg.

2 As the study is based on an online survey it was not possible to gain representative results for the German-speaking population that is older than 74 years.

2. Background and Relevance of the Study

2.1 Terms and Definitions

According to Bieger (2005), a destination is a geographical area (place, region, hamlet), which has been chosen as a travel destination by the guest (or a guest segment). It includes all the accommodation, catering and entertainment facilities necessary for a stay. For this reason it is the competitive unit in incoming tourism and must be run as a strategic business unit.

Consequently, the definition of the destination must be from the perspective of the guest, so that the significance of a strong customer focus in the modern marketing of a destination is emphasized. In addition to that a coordinated strategic management of the destination is essential.

The term "theme competence" used in the survey refers to theme marketing, which is widely established and practiced in German tourist organisations. Theme marketing in this context means the strategic marketing approach of the destination to accentuate themes within its communication policy such as types of holiday and holiday activities.

In order to be able to position the product mix developed by the destination successfully on the market, three core requirements have to be created: Firstly, a product mix has to be provided and coordinated; secondly, this has to be continuously adapted to dynamic market conditions, and finally, the product mix has to be communicated (Wöhler, 1997). The theme marketing of the destination in particular, plays a role here. The tourist target groups must be aware of the product mix available at the destination and be convinced that this will meet their needs at that time. The product mix has to be communicated to the target groups accordingly in order to generate tourist demand (Eisenstein, 2014). With the help of the theme marketing of the destination, and to meet the communication challenges, the individual components of the product mix of the destination are pooled in "themes". The two main objectives here are the emphasis on the core competencies of the destination and the differentiation between competitors. The choice of themes is usually made according to the ability and willingness of the destination to provide a corresponding product mix. Consequently, the approach focuses primarily on the prospective range of products offered.

In the "Destination Brand" study, the term theme competence refers to the eligibility of chosen themes in relation to the destination (types of holiday, holiday activities) from the viewpoint of the German population. It indicates how suitable the respondents think the respective destination is for them in terms of a particular activity (e.g. hiking) or a certain type of holiday (e.g. city break).

The theme competence of the destination can therefore provide important insights into the evaluation of the themes chosen by the destination. Subsequently, theme competence is crucial when making a decision regarding the financial controlling and further development of the theme package defined by the destination as part of its communication policy.

In addition to the theme marketing approach demonstrated, lately the conventional approach of target group marketing in the consumer market has been a source of discussion on the German destination market and has been introduced in many places. In this respect the demand perspective of destination marketing has been increasingly emphasized. As a reaction to the transformation of the tourist market into a buyer's market (see Section 2.2), the intensified significance of the demand side fulfils an important requirement in establishing more effective structures in the destination and is among the core characteristics of modern destination management (Eisenstein, 2014).

Although the inclusion of both the supply and demand perspective for the business segment cut off in strategic marketing has long been undisputed (Meffert, Burmann & Kirchgeorg, 2008) this data-based and conceptually developed approach has only been applied in the marketing of German destinations in the recent past. Meanwhile there is wide consensus among those working in destination management that a combination of both approaches, i.e. a combined view of the supply and demand perspective in terms of target group oriented marketing, is the most efficient way of marketing a destination. In target group-oriented marketing, product mix based on the core competencies of the destination, developed by the destination and communicated via themes, is allocated to high potential target groups.

2.2 Relevance of the Study

Not only the market situation, but along with it the competitive conditions for destinations have also changed drastically since the 1990s. In the course of globalisation, once existing barriers to exploring more distant destinations have been increasingly overcome due to improved transportation and information technologies ("Death of Distance"; Cairncross, 1997). Examples include the developments in air travel which have led to markedly cheaper offers on long-haul flights, the facilitation of monetary transactions, and the easier access to information via new media. The range of holiday destinations on offer and the mobility options available to the tourist consumer are continuing to grow, while in comparison, the growth of the (domestic) demand side is lagging behind (Eisenstein, 2014; Tschurtschenthaler, 1999).

The market entry of numerous new destinations has considerably intensified the competitive situation for traditional destinations (Bieger & Weibel, 1998). The market for holiday destinations has changed from a seller's market to a buyer's market, in which the wishes of the buyer dictate the product development (Eisenstein, 2014).

On the demand side, the growing travel experience of a wider social stratum in Germany has led to more specific market knowledge and at the same time to a reduction in travel risk perception (Pechlaner & Weiermair, 1999) as well as to a quantitative and qualitative inflation of expectations (Eisenstein 2014).

In addition to the developments in the tourist market described here, the possibilities of market cultivation are influenced by the fact that public means of financing German tourist organisations are becoming scarcer. Therefore, tourist organisations are under increasing pressure to legitimize their work or find themselves accountable for using governmental funding.

The changing basic conditions for the destinations and the problems arising from this for traditional destinations have led to a heightened awareness regarding the necessity of a strategic marketing basis in order to secure the long-term competitiveness of the destination (Eisenstein, 2014; Lewis-Cameron & Roberts, 2010; Ruhanen, 2007). This entailed a reassessment of the previous use of resources with the objective of making marketing as focused and efficient as possible.

Target group orientation and market segmentation have become increasingly important in destination management. The aim is to identify potential customer segments which are taking advantage of, or want to take advantage of the services of the destination in question, in order to satisfy their needs (Matzler & Pechlaner, 1999). Individual market segments can be served in a more effective and profitable way with systematic customer orientation than if the whole market was treated uniformly. The main advantages of market segmentation are the possibility of creating products in a target group-oriented manner, and the choice of customized sales channels and communication media (Kotler, Keller & Bliemel, 2007).

In order to remain competitive despite the quantitative and qualitative inflation of demand, the destination has to offer the target group a comprehensive product range (plurality of choice), which at the same time guarantees that the needs and increasing demand for quality will be satisfied for all the components of the product mix in the service supply chain (Eisenstein, 2014; Lewis-Cameron & Roberts, 2010).

At the same time however, the creative options in terms of the product mix of the destination are limited. The variation of the product mix of the destination is restricted due to the available production factors.[3] For this reason, core competencies among other things, must be defined for the destination according to the existing input factors (Figure 1).

Subsequently, potential market segments must be defined for these core competencies in order to attain a correlation between core competencies and market segment or a combination of product mix and target group. Firstly, one must bear in mind that the defined core competencies allow the potential target group to differentiate between competing destinations so the defined market segments become aware of their attraction (Matzler & Pechlaner, 1999). Moreover, the potential target groups must be big enough.

Figure 1: Approach to definition of themes for the marketing of the destination[4]

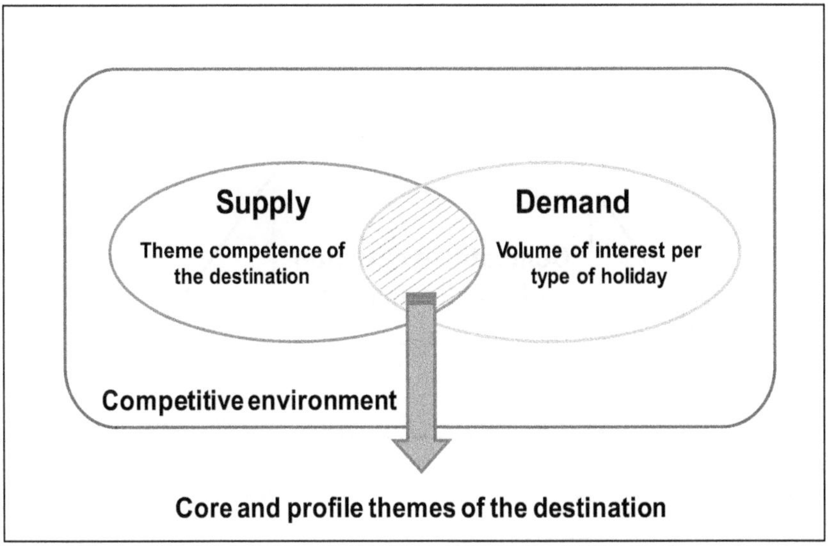

Thus, "Destination Brand" analyses the general potential of prospective customers among the German population in selected holiday themes to identify possible customer segments for the participating destinations (volume of demand; Figure 2).

3 On the different input and production factors in destinations see Ritchie & Crouch (2003) and Fischer (2009).

4 Source: IMT (2010).

By rating the assigned theme competence of each destination, this study can provide important indications on how to identify core competencies of a destination. These can then be used to create an appropriate product mix. The identification of core competencies takes place on the demand side whereby the attribution of eligibility for each theme and destination is determined from the viewpoint of the German population (eligibility rating of the product range by demand). Comparing the data of the various destinations provides insights into the relative position of the destination in the competitive environment (competition) with regard to the potential core competencies.

Figure 2: Competitive benchmark per touristic marketing theme[5]

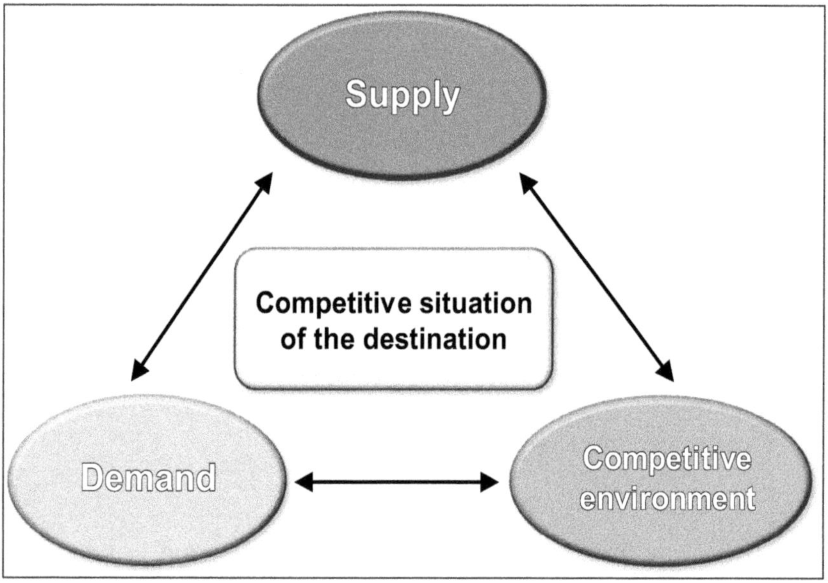

Taking into consideration the respective competitive position, the most promising product mix – target group combinations of the destinations can be determined by means of target group analysis of individual destinations. In this process, the strategic decisions of the target group-oriented marketing of the destination will be secured in a database. This is illustrated in the following example of the Ruhr Basin as a destination.

5 Source: IMT (2010).

3. "Destination Brand" – The Example of the Ruhr Basin Destination

3.1 Methodology

In October 2010, a total of 8,200 respondents, forming a representative cross-section of the German population aged between 14 and 74, were interviewed online. Various partial samples were used, each representative in its own right. With reference to the theme eligibility, the sample size for each of the 122 analysed destinations was at least 400, but in most cases the sample size was 1,000 respondents. For the assortment of the sample, methodical quota sampling was used.

The 122 travel destinations included the German federal states, selected cities and travel regions, and also communes with a high number of documented overnight stays by domestic travellers.

A total of 28 themes (holiday types/activities) were included in the study. These can be divided into five general themes and 23 special themes. The general themes "health vacation/journey (excluding "cure")", "culinary travel", "cultural travel", "outdoor vacation" and "spa break" were asked about for all of the travel destinations analysed. A further five of the 23 special themes were chosen for each destination. The selection of these special themes was conducted partly by the destination management organisations that bought the data or on the basis of research on the communication strategies of the destinations carried out by the Institute of Management and Tourism. For the Ruhr Basin travel destination, the selected themes included "attending events", "visiting industrial heritage sites", "biking (not mountain biking)", "shopping" and "city breaks".

In 2013, the study was repeated based on the same methodology. The number of analysed travel destinations rose to 137 German and 33 European travel destinations. The amount of examined themes increased to 55 while the general themes stayed the same as in the primarily study.

3.2 Derivation of Product-mix/Target Group Combinations Exemplified by the Ruhr Basin

3.2.1 *Volume of Interest (2010)*

First, the general volume of interest of the German population was determined for all themes, independent of the destination.[6] In view of the 28 themes analysed

6 The volume of interest is the result of the proportion of the top-two box values measured on a scale from "5=very interested" to "1=not at all interested".

in 2010, it is evident that the general German population were most interested in the following types of holiday/holiday activities: “Outdoor leisure” (70%), “city breaks” (68%) and “vacation by the sea/beach holidays” (64%)[7], followed by “appreciating culinary specialties” (54%) “spa breaks” (50%), “visiting cultural attractions / making use of cultural facilities” (50%) and “biking (excluding mountain biking)” (38%). In each case thirty-six percent of the German population are interested in “making use of health products and services”, “shopping” and “events” respectively. Of the ten themes offered in the Ruhr Basin portfolio, the least interest was shown in the “visiting industrial heritage sites” theme (26%).

Table 1 presents the findings for all holiday themes analysed in 2010, highlighting the theme portfolio of the Ruhr Basin.

Table 1: Competitive benchmark per touristic marketing theme[8]

	Volume of interest per type of holiday / activity ➢ top-two-box "5 = very high interest" + "value = 4"	% of cases	Projection*
1	Being outside in the nature	70%	40.9 million
2	City break	68%	40.1 million
3	Vacation by the sea / beach holiday	64%	37.5 million
4	Appreciating culinary / gastronomic specialties	54%	31.5 million
5	Applying wellness offers	50%	29.3 million
6	Visiting cultural institutions / applying cultural offers	50%	29.2 million
7	Family vacation	49%	29.1 million
8	Visitation of gardens / parks	48%	28.5 million
9	Hiking	47%	27.8 million
10	Biking (not mountain biking)	38%	22.5 million
11	Applying health offers	36%	21.4 million
12	Applying shopping possibilities	36%	21.2 million
13	Visitation of amusement parks	36%	21.2 million
14	Attendance of events	36%	21.1 million

	Volume of interest per type of holiday / activity ➢ top-two-box "5 = very high interest" + "value = 4"	% of cases	Projection*
15	Experiencing cultural circles / night life	36%	21.1 million
16	Luxury journey / vacation	35%	20.4 million
17	Shore leave vacation / farm holiday / vineyard holiday	29%	17.3 million
18	Visitating offers of industrial culture	26%	15.3 million
19	Wine tour	26%	15.0 million
20	Practicing water sports (not sailing)	26%	15.0 million
21	Practicing winter sports (cross-country skiing etc.)	25%	14.7 million
22	Sport vacation	25%	14.5 million
23	Nordic walking	16%	9.5 million
24	Mountain biking	14%	8.5 million
25	Film tourism (e.g. visitation of filming spots)	14%	8.1 million
26	Sailing	12%	7.3 million
27	Horse-riding	11%	6.5 million
28	Playing golf	6%	3.6 million

Ranking of the types of holiday / activity according to their level of interest

* Projection of the absolute volume of interest (number of persons).

---

7 Accordingly comparing the analysed themes of the portfolio of the Ruhr Basin the holiday activities “outdoor leisure” and “city break” by far have the highest general volume of interest, independent of the destination.

8 Source: IMT (2010).

3.2.2 The Attribution of Theme Competence 2010

A core element of the study was to ascertain how suitable the German population finds the destinations in question in terms of types of holidays and holiday activities.

Grounded Theme Competence

Grounded theme competence is based on how eligible the respondents that are familiar with the destination or know it at least by name think the destination is for the themes analysed.[9] In contrast to the as well analysed "first comes to mind" theme eligibility, which will be introduced later in the text, regarding the grounded theme competence the respondents evaluated the competence of given destinations.

A representative sample of the German population thinks that of the ten themes considered for the Ruhr Basin, the "visits to industrial heritage sites" theme is the one that best represents the theme competence of the Ruhr Basin (61%). This is closely followed by "city breaks" (54%), "events" (52%), "shopping" (48%) and "cultural travel" (44%). Twenty-nine percent and 27% consider the region suitable for "biking" or "culinary travel", respectively. Not many consider the Ruhr Basin suitable for the themes "spa break" (20%), "outdoor vacation" (15%) and "health vacation" (9%) (Figure 3).

In addition it was analysed if the Ruhr Basin is eligible for the following relevant subgroups in terms of the theme portfolio:

- "People interested in the themes": have a general interest in the type of holiday or activity[10]
- "Previous visitors": have already spent at least one night in the Ruhr Basin
- "Previous visitors who are interested in the themes": have a general interest in the type of holiday or activity and have already spent at least one night in the Ruhr Basin

The Ruhr Basin obtained very high approval ratings for the "industrial heritage theme" from the subgroup of "People interested in the themes" (71%), "Previous visitors" (86%) and "Previous visitors who are interested in the themes" (93%). It is shown that in this case the higher the involvement in the theme or destination, the higher the approval rating.

---

9 The eligibility is measured by the proportion of the top-two box values measured on a scale of "5 = very suitable" to "1 = not at all suitable".

10 The general interest is defined as the proportion of the top-two box values measured on a scale of "5=very interested" to "1=not at all interested".

Figure 3: Theme competence of the Ruhr Basin[11]

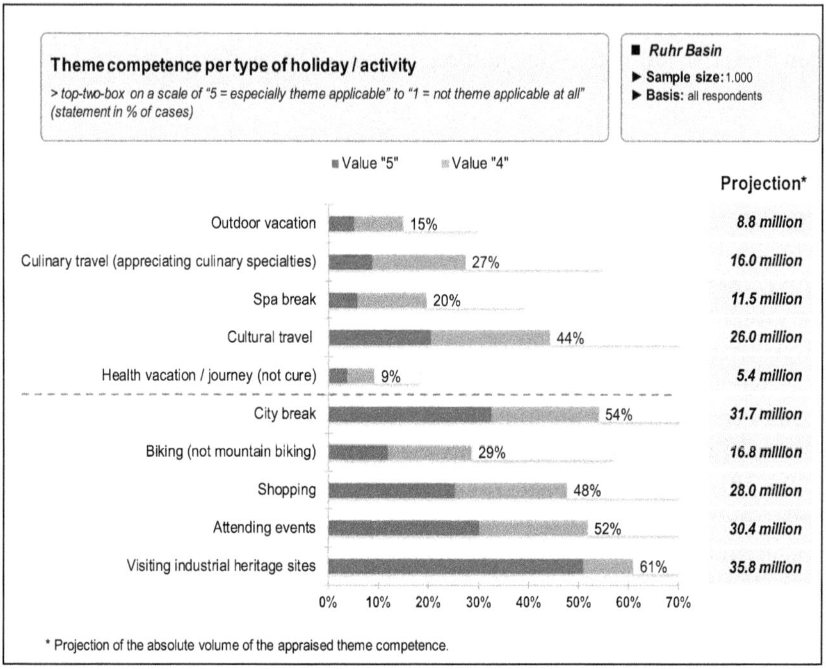

In the course of a target group analysis for the Ruhr Basin, the eligibility of the themes was analysed, differentiated by the target groups as defined by the destination:

- "New families" (26–55 years old; domestic net income: above 1,500 Euro/month; household with at least one resident child under 14 years of age)
- "Young adults" (under 35 years old; 1- or 2-person household; no resident children under 14)
- "Well-educated best ager" (48–74 years old; education: A-levels or university degree)

The aim of the target group analysis was to identify the themes with the greatest potential for the corresponding target group in order to be able to meet demand in a better way. Using the Kruskall Wallis test, significant differences were

11 Source: IMT (2010).

found in the "outdoor vacation", "visiting industrial heritage site" und "biking" themes.[12]

For the "outdoor vacation" theme, the approval ratings by the "well-educated best agers" target group were well above average (20%), while the approval rating by "young adults" (10%) and "new families" (11%) was slightly below the total average (15%).

For the "visiting to industrial heritage sites" theme, the Ruhr Basin received a high approval rating from the "well-educated best agers" (70%), while the "young adults" (60%) and the "new families" (63%) rated this theme slightly below and slightly above the total average of all respondents (61%), respectively.

The approval rating from the "well-educated best agers" for the "biking" theme was also well above average (41%). The "young adults" do not consider the Ruhr Basin very suitable for this type of holiday activity (25%), while the approval rating of the "new families" (31%) is slightly above the total average of all respondents (29%).

Theme Eligibility of "First Comes to Mind"

In addition to the grounded theme competence, an analysis on the "first comes to mind theme eligibility" was conducted. In this case the respondents were asked to imagine that they wanted to go on a special theme holiday. Without being given pre-defined responses, they were asked which holiday destinations in Germany they would consider suitable for each of the themes they had chosen. Up to three answers could be given here. This question was asked for all themes that were analysed.

With regard to the ten themes relevant for the Ruhr Basin destination, the "visiting industrial heritage sites" theme clearly stands out. In 54.9% of the cases, the Ruhr Basin was associated with being especially suitable for an "industrial heritage" trip. The frequent mention of the Ruhr Basin during the "first comes to mind" survey on the "industrial heritage" theme makes this theme more distinctive for the region and thus makes it possible to differentiate it from its competitors. The results the Ruhr Basin received for the "cultural travel" (7.2%) and "attending events" (7.3%) are also worth mentioning as in comparison to the competitors the scores are still rather high. Even the "shopping"

---

12 The definitions of the target groups overlap in some criteria. Therefore the test for the statisticalsignificance (Kruskall Wallis test) could only be carried out with regard to the question if significant differences between each "counter group" could be demonstrated e.g. between the "new families" and the "households with no resident children under 14".

theme was considered especially suitable by 2.3% of the respondents while only 1.3% thought the same for the "biking" theme. The Ruhr Basin attained a share of under 1% for all other themes in the portfolio.

3.2.3 Competitive Environment (2010)

Furthermore, competition benchmarking was carried out for each of the destinations analysed. This involved a comparison with the competitors in terms of the "first comes to mind theme eligibility" and the "grounded theme eligibility".[13]

The competition benchmark for the "first comes to mind" question ranked the Ruhr Basin in first place in the "industrial heritage sites" type of holiday. This corroborates the conclusion that the "industrial heritage" theme is a high profile force for the region. The Ruhr Basin ranked fifth place for "travelling to events" and seventh place for "cultural travel". The "shopping" theme ranked tenth, while the other themes did not make it to the top ten.

A comparison with all other travel destinations was made for the Ruhr Basin during the competition benchmarking on "grounded theme competence", based on the top-two box values. Furthermore, in order to define the competitive environment more clearly for the destination in question, further competition benchmark analyses were carried out for the destination in selected categories, e.g. "general region". The competition benchmark for each destination category for the themes in the Ruhr Basin portfolio in question contains the ranking of the competing destinations and the mathematic average value of the destination category.

A total of 122 travel destinations were considered during the competitive benchmarking regarding the general themes. The Ruhr Basin obtained a very low competitive ranking for the following themes: "health vacation (excluding cure)" (104th place), "outdoor vacation" (100th place) and "spa break" (92nd place). The ranking for the "culinary travel" theme was average (61st place). Out of all general themes, the Ruhr Basin only ranked in the top 20 for the "cultural travel" theme (15th place).

With view to the special themes of the Ruhr Basin the number of destinations analysed, and thus the comparative basis, varies considerably depending on the choice of the destination management organisation or the research-based

---

13 In addition to the competitive benchmark based on all respondents, the analyses were carried out on the basis of the subgroup of the "People interested in the themes".

specification by the IMT. The Ruhr Basin ranks among the top destinations for "attending events" (8th place out of 33) "shopping" (7th place out of 26), and "city breaks" (15th place out of 45), while it obtains an average ranking for the "biking" theme (58th place out of 97). Particular mention should be made here of the Ruhr Basin topping the other examined destinations by far for the "visiting industrial heritage sites" theme and thus ranking 1st place, although compared to the other themes, the lower degree of competition with only ten competitive destinations analysed must be considered here (Figure 4).

Figure 4: Competitive benchmark per touristic marketing theme[14]

Category: All surveyed destinations per touristic marketing theme | Basis: all respondents

Ruhr Basin (n = 1.000)	Own value	Ø of the category	Number of destinations in the category	Ranking
▪ Health vacation / journey (not cure)	9%	32%		104.
▪ Culinary journey (appreciate gastronomic offers)	27%	28%		61.
▪ Cultural journey / vacation	44%	26%	122	15.
▪ Natural vacation	15%	39%		100.
▪ Wellness vacation / journey	20%	32%		92.
▪ Attendance of events	52%	35%	33	8.
▪ Visitation of offers of industrial culture	61%	28%	11	1.
▪ Biking (not mountain biking)	29%	33%	97	58.
▪ Shopping	48%	38%	26	7.
▪ City break	54%	41%	45	15.

3.2.4 Combination of the Three Dimensions

In conclusion, using the quadrant analysis chart shown in Figure 5, the core findings of the theme competencies of the Ruhr Basin were summarized visually in a three dimensional analysis of the competitive situation.

The three envisaged dimensions of analysis are accounted for in the quadrant analysis as follows:

14 Source: IMT (2010).

- “Demand”: The X-axis shows the general potential interest in each theme. The greater the interest shown by the German population in the respective holidays, the further right the theme is positioned in the quadrant.
- “Supply”: The theme eligibility assigned to the Ruhr Basin is illustrated by the size of the respective “theme spheres” The bigger the respective “theme sphere” is, the more eligible the respondents found the Ruhr Basin for the theme in question.
- “Competition”: The relative position of the Ruhr Basin in the competitive environment among all the destinations analysed per theme is shown on the Y-axis. The better the Ruhr Basin performs per theme in the competitive benchmark, the higher the respective themes are positioned in the quadrant.

The quadrant allows for the overall content assessment of the themes in the Ruhr Basin portfolio. Its modus operandi is illustrated using the “outdoor vacation” and “industrial heritage” themes as examples:

The “outdoor vacation” theme stands out due to its high potential of general interest. However, for the “outdoor vacation” theme, the Ruhr Basin was assigned very low theme eligibility (small “theme sphere”). Its relative position in the competitive environment is also very low. On the other hand, the general potential interest of the German population in the “industrial heritage” theme is considerably less pronounced. However, the Ruhr Basin attained a very high eligibility rating for this theme (large “theme sphere”) as well as a very strong relative position in the competitive environment.

Figure 5: Three-dimensional analysis of the competitive situation[15]

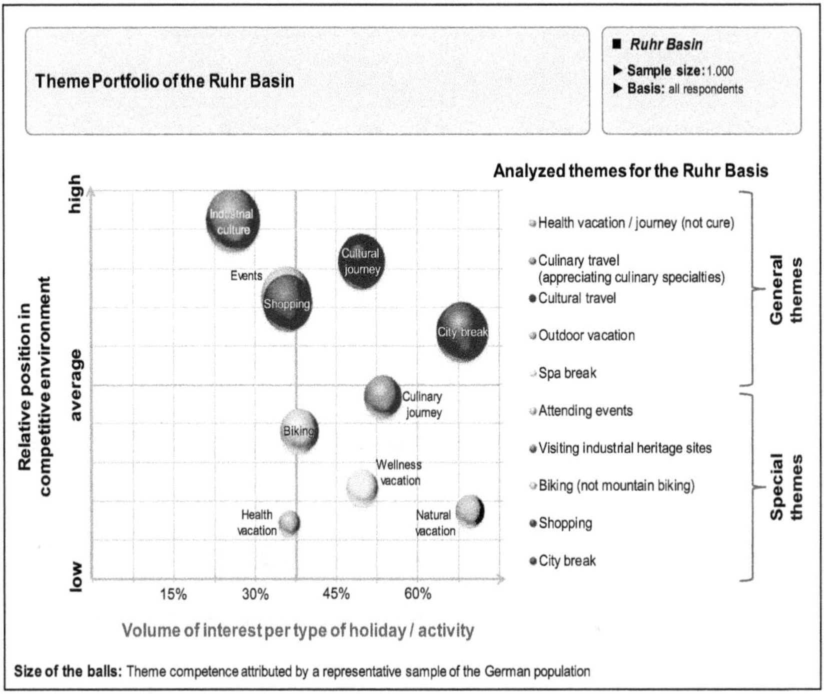

The results of the three-dimensional competitive benchmark shown here for all the respondents were also analysed for the three designated target groups of the Ruhr Basin. Thus a well-grounded basis for the definition of product mix – target group combinations for the destination as well as for the development of target group-oriented theme marketing is given. Figure 6 provides a summary overview of this.

15 Source: IMT (2010).

Figure 6: Target group-check for the Ruhr Basin[16]

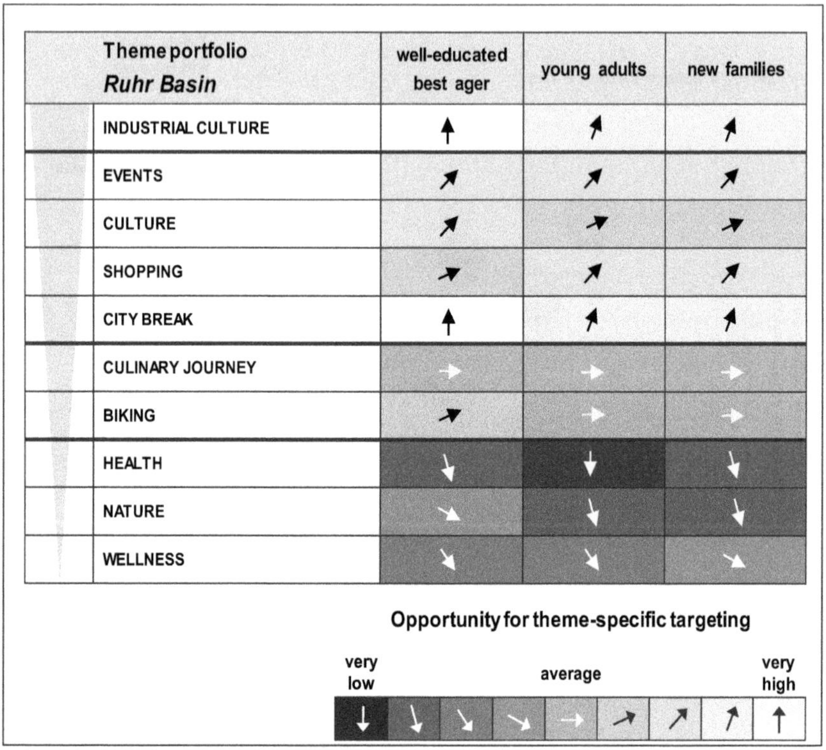

3.2.5 *Repetition of the Study in 2013 – Further Insights Based on a Time Comparison*

As already mentioned, the "Destination Brand" study was carried out again three years later, based on the same methodology. Therefore, the development of the volume of interest per type of holiday / activity among the German population, as well as changes in the assessment of theme competence and the competitive situation of the analysed travel destinations can be monitored as well.

For the Ruhr Basin, and the other participating travel destinations, the time comparison gives further insights for strategic decisions. For example, it shows whether the touristic theme represents a growing or decreasing market in Germany with view to the volume of interest. In concrete terms, the volume of

16 Source: IMT (2010).

interest in "city breaks" and "visiting industrial heritage sites" has decreased from 2010 to 2013 by three percent points, while "attending events" (+4 percent points) and "biking (not mountain biking)" (+ 2 percent points) show a small growth of volume of interest. Despite this slight decrease, the total volume of interest in "city breaks" is still very high (65%).

Figure 7 presents the time comparison of the volume of interest for the five selected special themes of the Ruhr Basin.

Figure 7: Volume of interest per type of holiday / activity – time comparison[17]

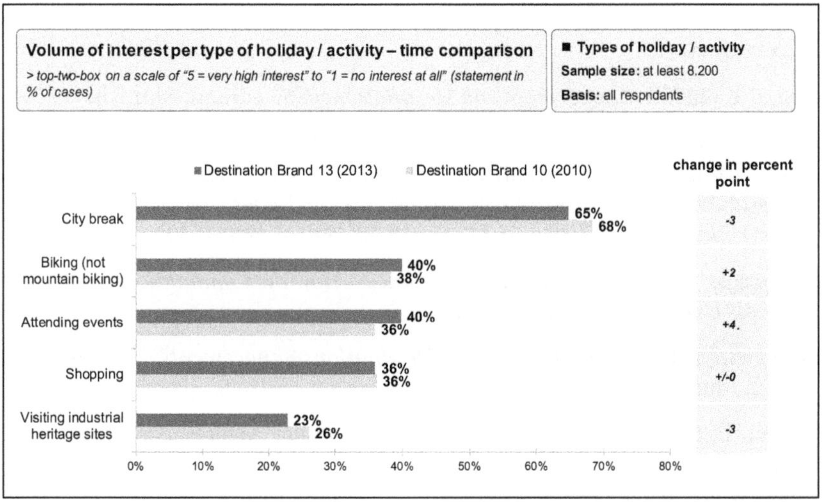

4. Conclusion and Outlook

Due to changing market conditions, the strategic management of German travel destinations has gained in significance in professional destination management in the last few years. The example of the Ruhr Basin destination demonstrated that the "Destination Brand" study can contribute to data-based decision-making with regard to the defined product mix – target group combinations of German travel destinations.

In the course of further research, with the help of the data accumulated in the course of the study, and using multivariate methods of analysis, the interrelation between various themes will be analysed in order to determine complete

17 Source: IMT (2010).

theme packages, the potential interest in these packages and their potential target groups.

This further research approach allows for the practice-oriented notion that it is not always the fulfilment of a single need that is decisive for the consumer, but rather the fulfilment of a needs package. The findings will allow the allocation of various theme combinations to potential target groups. Thus, in addition to the findings presented here, the "Destination Brand" study can provide information with decision relevance for the strategic development of destinations.

References

Bieger, T. (2005): *Management von Destinationen.* 6th edition. Munich.

Bieger, T./Weibel, C. (1998): Möglichkeiten und Grenzen des kooperativen Tourismusmarketing. Schaffung von Tourismussystemen als Strategie gegen destinationsähnliche Konkurrenzprodukte. In: Keller, P. [Ed.]: *Destinationsmarketing. Möglichkeiten und Grenzen.* St. Gallen, p. 167–200.

Cairncross, F. (1997). *The Death of Distance – How the Communication Revolution Is Changing Our Lives.* London.

Eisenstein, B. (2014): *Grundlagen des Destinationsmanagements. 2nd revised edition.* Munich.

Fischer, E. (2009): *Das kompetenzorientierte Management touristischer Destinationen – Identifikation und Entwicklung kooperativer Kernkompetenzen.* Wiesbaden.

Institut für Management und Tourismus (IMT) (2013): *Destination Brand 13. Themenkompetenz deutscher Reiseziele.* Heide / Holstein.

Institut für Management und Tourismus (IMT) (2010): *Destination Brand 10. Themenkompetenz deutscher Reiseziele.* Heide / Holstein.

Kotler, P. / Lane Keller, K. / Bliemel, F. (2007): *Marketing-Management. Strategien für wertschaffendes Handeln.* 12th revised edition. Munich.

Lewis – Cameron, A. / Roberts, S. (2010): Strategic Destination Marketing: The Key to a Competitive Advantage. In: Lewis-Cameron, A. [Ed.]: *Marketing island destinations: concepts and cases.* Amsterdam, p. 11–22.

Matzler, K. / Pechlaner, H. (1999): Kompetenzorientierte Entwicklung von Kundenaquisitionsstrategien. In: Pechlaner, H. / Weiermair, K. [Eds.]: *Destinations-Management. Führung und Vermarktung von touristischen Zielgebieten.* Wien, p. 137–158.

Meffert, H. / Burmann, C. / Kirchgeorg, M. (2008): *Marketing. Grundlagen marktorientierter Unternehmensführung. Konzepte – Instrumente – Praxisbeispiele.* 10th completely revised edition. Wiesbaden.

Pechlaner, H. / Weiermair, K. (1999): Neue Qualifikationsanforderungen in Destinationsorganisationen. In: Pechlaner, H. / Weiermair, K. [Eds.]: *Destinations-Management – Führung und Vermarktung von touristischen Zielgebieten.* Wien, p. 79–90.

Ritchie, J.R.B. / Crouch, G.I. (2003): *The Competitive Destination – A sustainable Tourism Perspective.* Cambridge.

Ruhanen, L. (2007): Destination Competitiveness: Meeting Sustainability Objectives Through Strategic Planning and Visioning. In: Matias, A. / Nijkamp P. / Neto P. [Eds.]: *Advances in Modern Tourism Research. Economic Perspectives.* Heidelberg, p. 133–152.

Tschurtschenthaler, P. (1999): Destination Management / Marketing als vorläufiger Endpunkt der Diskussion der vergangenen Jahre im alpinen Tourismus. In: Pechlaner, H. / Weiermair, K. [Eds.]: *Destinations-Management. Führung und Vermarktung von touristischen Zielgebieten.* Wien, p. 7–35.

Wöhler, Kh. (1997): *Marktorientiertes Tourismusmanagement – Tourismusorte: Leitbild, Nachfrage- und Konkurrenzanalyse.* Berlin.

Eva Holmberg and Kaija Lindroth

The challenges of destination branding – a case study of Helsinki

1. Introduction

Branding has over the recent years been regarded as a strategy to gain competitive advantage, initially for organizations, but later on also for nations and tourism destinations. Thus the branding of places is part of the discourse on how nations and regions can improve their competitiveness in a global market (Hall 2008).

Pike (2009) identifies several research gaps related to branding destinations in scientific articles, for example, the lack of in depth case studies of the politics of destination brand decision making. Also Boo, Busser & Balagou (2009) stress the fact that most work has focused on measuring the brand image of destinations rather than on deepening the understanding of the brand building process itself. This study strives to enhance knowledge about these gaps.

The aim of this study is to explore and critically evaluate the brand building process in the Helsinki metropolitan area, from its start in 2004 to the present day. The main focus will be on the two main actors in the process, the Helsinki City Tourism and Convention Bureau (HTCB) and the Greater Helsinki Promotion Ltd (GHP), and their actions. The former works for promoting Helsinki among international tourists and the latter for bringing international investments into the capital area. This study focuses on the branding process itself instead of assessing the actual brand identity or the value of the brand. The view taken in the study is that of the owners of the process. The awareness of Helsinki as a tourism destination grew in the 1960s but Helsinki is still a rather small player in the international tourism market. Thus, Helsinki has rarely been the focus of academic tourism research, even though its branding strategy has been discussed, for instance by Hospers (2011).

2. Brands, destination brands and the brand building process

2.1 Brands in general

The concept of branding has long roots. Already in the Stone Age hunters used weapons of certain "brands" to succeed, but it was in the 16th century when the idea of brand as we know it today started to emerge. In the 1950s the expansion of branding theory started due to the overall development of mass media. (Hampf and Lindberg-Repo 2011)

Brands can be seen as symbols around which social actors such as firms, organisations, customers as well as nations construct identities (Pitt, Opoku, Hultman, Abratt and Spyropoulou 2007). Aaker's (1991, 7) definition of a brand as "a name, term, sign, symbol or design, or a combination of them intended to identify the goods and services of one seller or group of sellers and to differentiate them from those of competition" is the most cited in literature (e.g. Kotler and Gartner 2002, Pike 2010). On the other hand, it is also stressed that a brand is more than a slogan, symbol or promotion (Therklesen and Gram 2010). In fact all brands are products to begin with, but through creativity or extraordinary value the product has been developed into a brand (Cai and Hobson 2004).

Brands differentiate products (Pike 2009) and represent a promise of value (Kotler and Gartner 2002). For consumers brands have both social and emotional value; many consumers are thus willing to pay a premium price for a brand they prefer (Kotler and Gartner 2002). The importance of having a brand is related to brand loyalty, i.e. many consumers are reluctant to change a brand they really like and one that has been proven good (Cai and Hobson 2004).

Figure 1: Difference between brand identity and brand image (modified from Pike 2004)

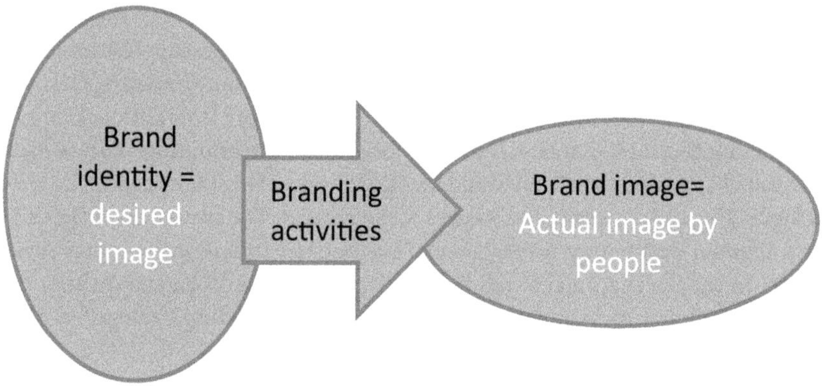

Figure 1 shows the key idea of branding. Brand identity is the foundation of any brand building strategy. The brand identity is the unique set of brand associations that a brand strategist aspires to create or maintain; it is the soul and vision of the brand (Aaker 1996, 68). Any company focusing on developing its brand, has to make sure that the whole organisation understands the brand identity and take responsibility for it in their daily work. Otherwise it is likely that the desired brand identity will not be communicated to existing and potential customers (Joachmisterhaler and Aaker 2002).

Branding activities such as designing product names, slogans and symbols will, together with for instance, word-of-mouth influence, shape the actual brand image people have of a product or an organization (Pike 2010). The brand image is thus the associations existing and potential customers have of a brand. The image can be positive or negative and it is usually developed in the mind of customers over a longer period of time (Anholt 2010). In order to develop stronger brand identities certain features, such as creating a human touch or a personality, are often stressed. Brand personality is defined by Aaker (1996, 141) as "the set of human characteristics associated with a brand". Both product and service brands are developed to include human characteristics, such as sporty, energetic or sophisticated (Usakli and Bologlu 2011).

Brand awareness is related to brand recognition and brand recall, both of which are dependent on whether a consumer is able to identify the brand among other similar brands. Even though the awareness of a brand may be high it does not automatically contribute to a positive association among the consumers. The image of the brand is more important than the overall awareness, i.e. a unique and positive brand image among the consumers will often result in brand loyalty (Cai and Hobson 2004).

The success of a brand is described by how much equity exists in the name of the brand, which is presented on the corporate balance sheet (Pike 2010). Brand equity is dependent on the awareness and the image of the brand. A positive and solid reputation will make up an significant intangible value for a company or organization (Anholt 2010). The brand equity can be described in financial terms or from a consumer-based perspective. If brand equity is studied from a financial perspective it is possible to distinguish between the total physical assets of a company and its immaterial brand assets (Hampf and Lindberg-Repo 2011).

In contrast to tangible products, services brands are personal rather than standardized, memorable rather than ordinary, and created rather than just made and delivered (Cai and Hobson 2004). Due to the fact that a service is intangible, produced and consumed at the same time, and of heterogeneous quality, a service brand requires a slightly different conceptualization compared to product branding (Garcia et al. 2011). While a product brand can be developed mainly through planned communication, a service brand is developed during the service process as well. Thus, a strong command over the service delivering process is the key to a strong brand in service companies. Poor service quality can easily destroy a brand (Moilanen and Rainisto 2009).

2.2 Branding places

According to Garcia et al. (2011) the conceptualization of product and service branding can be the starting point for the process of the branding of places, such as tourist destinations. Generally all knowledge of services and product branding is applicable to place branding. The fact that a geographical area is a diverse system comprising different actors such as companies, public organisations and local people, makes the branding of places a more complex and challenging process than branding a company, a product or a service. Moreover, most places have a shortage of marketing knowledge and expertise, which does not make the branding process any less challenging (Moilanen and Rainisto 2009).

Hall (2008) stresses that brand building is a crucial activity to improve the competitiveness of geographical areas. Whilst the definition given by the American Marketing Association is commonly accepted as a suitable definition of a product brand, it has been harder to agree on an appropriate definition of place brand (Pike 2009; Pitt et al. 2007). According to Hall (2008, 60), place branding is "the development of a place brand and its promotion in order to differentiate a place from other location so as to gain advantage for its firms, organizations, people, products and services."

Place branding can be divided into internal and external place branding. The aim of internal place branding is brand development in relation to place identity, comprising activities aiming at improving community pride and the creation of an attractive environment for living. External place branding is about communication of the brand to people and organizations outside the area (Hall 2008).

The image of a country is dependent on its history, geography, culture and famous people which influence the image of the country both negatively and positively. Sometimes country images are very stereotypical, derived from impressions rather than facts. Strategic place marketing, including branding strategies, can support and also strive for changing the image of a geographical area, such as a country or a region (Kotler and Gartner 2002; Anholt 2010). From a country perspective re-branding is necessary since circumstances change (Pitt et al. 2007). On the other hand the fundamental beliefs people have of certain countries are not easily changed by marketing communication alone (Anholt 2010). In order to be effective, an image must be authentic, believable, simple, appealing and distinctive (Kotler and Gartner 2002). The image can be improved by working on the brand identity of the country, which may be influenced by tourism promotion and peoples' own experiences, the exported products and services, the political decision-making, cultural exchange, and the people of the country themselves (Anholt 2010).

2.3 Branding tourism destinations

The idea of branding tourism destinations has been applied only for around a decade (Usakli and Bloglu 2011; Pike 2010). According to Tasci (2011), tourism destination branding is easier than branding a country, because the process includes dealing with relatively controllable factors, such as marketing to key markets. One of the main challenges related to destination marketing and branding is the fact that many destinations are very similar and offer basically the same tourism services and products to tourists (Moilanen and Rainisto 2009). A destination brand has been defined by Ritchie and Ritchie (1998, 103) as:

> A name, symbol, logo, word mark or other graphic that both identifies and differentiates the destination; furthermore, it conveys the promise of a set of memorable travel experience that is uniquely associated with the destination; it also serves to consolidate and reinforce the recollection of pleasurable memories and the destination experience.

A wide-ranging definition of the activity of destination branding is offered by Blain, Levy and Ritchie (2005, 337), who define destination branding as:

> A set of marketing activities that (1) support the creation of a name, symbol, logo, word mark or other graphic that readily identifies and differentiates a destination; that (2) consistently convey the expectation of a memorable travel experience that is uniquely associated with the destination; that (3) serve to consolidate and reinforce the emotional connection between the visitor and the destination; and that (4) reduce consumer search costs and perceived risk. Collectively, these activities serve to create a destination image that is positively influence consumer destination choice.

In order to develop all the important elements of the tourism system, destination management organisations have been founded at many destinations (Bornhorst, Ritchie and Sheehan 2010). If branding at country level is pursued by governments and other public organisations in order to attract investments and to upgrade the profile of the country abroad, destination management organisations (DMOs) are working on improving the attractiveness of their destination (Usakli and Bloglu 2011; Boo et al. 2009). The DMO has a key role in the brand building process, but local people and entrepreneurs are also important actors when the promise of the brand is to be fulfilled (Garcia, Gomez and Molina 2012; Blain et al. 2005).

Destination branding work is beneficial for both the supply and the demand side of tourism. Branding offers a possibility to differentiate the destination, thereby gaining competitive advantage in a tourism market characterised by fierce competition and many rather similar destinations (e.g. Pike

2009; Garcia et al. 2012; Usakli and Baloglu 2010). Moreover, branding makes premium pricing possible (Pitt et al. 2007; Blain et al. 2005) and can increase visitor loyalty, benefiting thereby all the companies representing the supply side of tourism (Pitt et. al 2007; Pike 2009). The tourists will also draw advantages of destinations' efforts to brand themselves since it will result in lower search costs, reduced risks and in enhanced "brag value" (Pike 2009; Blain et al. 2005).

Tourism destination image is dependent on intangible and tangible elements (Garcia et al. 2012; Boo et al. 2009), and was defined by Crompton in 1979 as "the sum of beliefs, ideas and impressions that a person has of a destination" (Jenkins 1999, 2). An understanding of what image of a destination potential visitors have is important, since images influence the decision-making and the levels of satisfaction of tourists who have visited an area. Thus DMOs often pay research organizations for conducting studies on the destination image in its key markets (Jenkins 1999). It must be stressed though, that image formation is not branding, neither is image building. The difference between image building and brand building is that in branding the focus lies on the creation of a unique brand identity (e.g. Boo et al. 2009).

2.4 The destination branding process

Branding a destination should be more than the development of logos and taglines used in promotion, letters and business cards (Blain et al 2005). Morgan and Pritchard (2005) highlight that a destination brand can be developed in many different ways, for instance through traditional promotion such as advertising and personal selling, but also by websites as well as through public relations and co-operation with media. Destination branding is described as challenging due to the complexity of the tourism system of a destination (Boo et al 2009). Through coordination between the tourism actors at the destination in work for a common vision, it is more likely that the high-quality experience promised by the brand is delivered to the tourists (Blain et al 2005; Moilanen and Rainisto 2009).

The tourism actors involved in the branding process are aall stakeholders, defined by Freeman (1984, 31–32) as "those groups without whose support the organization would cease to exist", and originally included shareowners, employees, customers, lenders and society. Stakeholders can be divided into primary and secondary stakeholders depending on their ability to co-operate rather than challenge the organisation. In the brand building process of a destination the primary stakeholders are strategic partners, but it can also be important to communicate with secondary stakeholders (Garcia et. al 2012). According to Gilmore (2007) and Anholt (2010), potential stakeholders to be involved in the brand building process include residents, visitors (tourists), tourism companies, other service companies, media, politicians and export purchasers.

Branding a tourism destination is not a one-time project; successful destination branding should rather be seen as an ongoing process. An example of how the brand building process may look like is presented in Figure 2.

Figure 2: A conceptual model for destination branding (modified from Moilanen and Rannisto 2009, 166)

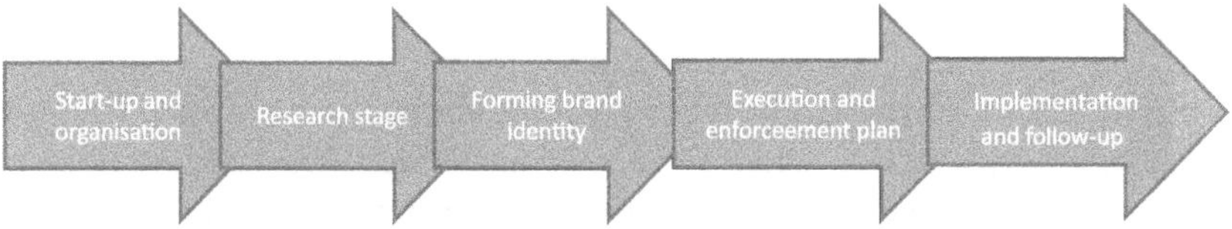

According to the destination branding model by Moilanen and Rannisto (2009) the process should comprise five stages:

1) Start-up and organisation stage, during which the commitment of key stakeholders should be secured, networks among the actors built, and the visibility of the project enhanced.
2) Research stage, during which the image of the destination is analysed (both externally and internally) by, for example, focus group interviews aiming at identifying the core identity of the brand.
3) Forming brand identity, during which stage the elements of the brand identity are chosen and tested and a strategic plan for communicating the brand prepared.
4) Execution and enforcement plan, during which operational plans for the brand are prepared.
5) Implementation and follow-up, when the brand is further tested, and development of service processes and physical infrastructure continue.

When it comes to the branding of destinations, the branding strategy must be based on the real attractors of the destination (Kotler and Gartner 2002). Moreover, the personality of the brand must be considered since this is a key factor to success (Usakli and Bloglu 2010; Pitt et al. 2007). Instead of selling beautiful beaches and friendly people, destinations should focus on highlighting the unique attributes of the destination, which are different from the attributes stressed by the main competitors (Garcia et al. 2012).

3. Methodology

This study employs a case-study approach. A case study is an empirical inquiry that investigates a contemporary phenomenon within its real-life context, especially when the boundaries between the phenomenon and the context are not clearly evident. A comprehensive understanding is reached through a process known as thick description, which involves an in-depth description of the entity being evaluated, the circumstances under which it is used, the characteristics of the people involved in it, and the nature of the community in which it is located (Yin 2003).

Case study research is recommended when research addresses the "how" and "why" questions (Yin 2003; Xiao and Smith 2006). Although case study research has been criticized for being unscientific and subjective, Xiao and Smith (2006) highlight that this criticism is not completely justified. Case studies can be seen to satisfy the three tenets of the qualitative method: describing, exploring, and explaining (Yin 2003).

Case studies can be either single or multiple-case designs. Single cases are used to confirm or challenge a theory, or to represent a unique or extreme case (Yin 2003). The data can stem from various collection methods, such as interviews, observations, documents and videotapes. The quality of a case study is guaranteed by different forms of triangulation such as theory, methodological and data source triangulation (Stake 1995; Guion 2002). This study relies mainly on data triangulation, i.e. multiple evidence in form of different types of data was collected and analysed.

This research employs a single case study, aiming at exploring the brand building process of Helsinki. In case study research cases are rarely chosen randomly, instead cases are selected based on interest and access. In this study, the researchers followed the brand building process of Helsinki from an outsider perspective for several years, and they knew the person having the strongest involvement in the destination branding process. The data collected and analysed comprises interviews with three experts strongly involved in building the brand of Helsinki, as well as several articles and company reports.

In order to guarantee the best possible quality of the study a focus was placed on triangulation in order to get a thick description of the brand building process of Helsinki. Multiple forms of evidence, such as documents, interviews, and articles were collected and analysed in order to gain a deep understanding of how different actors have worked for developing the Helsinki brand.

4. Results

4.1 Case: Branding Helsinki

Helsinki as a Tourist Destination

Helsinki, the capital of Finland, together with the surrounding municipalities (Helsinki region), is the home of 1.3 million people. It was founded in 1550, and received capital status in 1812 (HTCB 2014). Helsinki has never been a major tourism destination for leisure travel on a global scale. From a Finnish perspective though, Helsinki is the leading destination both for domestic and international tourism (Vuoristo 2002). Possessing attractions, such as the amusement park Linnanmäki, Lutheran and Orthodox churches, and the UNESCO World Heritage Site Suomenlinna, Helsinki offers several attractions for domestic tourism, as well as for tourists from the neighbouring countries such as Russia, Estonia and Sweden, which generate most of the visitors. Around 15 million tourists visited Helsinki in 2011, contributing to 3.2 million bed nights. Helsinki is a successful meetings venue, and attracts a large number of cruise tourists (Halonen 2011a).

4.2 Actors involved in branding Helsinki

The Helsinki Tourism and Convention Bureau (HTCB)

Helsinki's destination management organisation, the Helsinki Tourism and Convention Bureau (HTCB), was founded in 1963. It promotes Helsinki as a tourist and meeting venue, informs of travel services in the region, enhances the development of the tourism industry, and strengthens the appeal of the city. Its vision is to be the engine of the tourism development in the Helsinki region, fuelling the controlled and sustainable growth of the tourism industry (Halonen 2011b).

Greater Helsinki Promotion (GHP)

Greater Helsinki Promotion (GHP) is the investment promotion agency for the Greater Helsinki area. The agency helps international companies to create and conduct business in the Helsinki area. It is a non-profit organisation, established by the cities of the capital region to attract strategic international investments to the region. Its main responsibilities are consultation to companies interested in investing in Helsinki area, development of key investment drivers of the area, and strengthening the international identity and positive reputation of the Helsinki area (GPH 2011a).

According to the manager of the organisation, Tatu Laurila (2011), the work of GHP is not at all related to tourism. The organisation strives for unifying the metropolitan area of Helsinki, which comprises Finland's three biggest cities: Helsinki, Espoo and Vantaa. For success it would be important to involve all critical stakeholders in co-operation and in social capital sharing, but this is still a in progress. The cities financing the agency are also in many ways competitors, for instance when it comes to public capital, tax payers, and companies creating employment.

4.3 The Brand Building Process of Helsinki

This next section will provide an overview of the process of improving the brand of Helsinki in international markets, which mostly has taken place as two parallel processes, the first related to the tourism brand and the second related to branding Helsinki as a business location. At some stages the stakeholders have though successfully worked on common issues.

Branding Helsinki for tourists

Prior to the 1990s, Helsinki as a tourism destination was mainly a city attracting small amounts of tourists from the countries nearby, such as Sweden and Germany. The border to the East European countries was more or less closed, but since the breakdown of the Soviet Union in 1991 the inhabitants of countries such as Russia, Estonia and Poland have been allowed to travel freely. The opening of the borders resulted in a large growth of tourist numbers in Helsinki, especially from Russia and Estonia. During this time, in the early 1990s, the city of Helsinki tried to develop its identity as a modern European city in the international tourism market (Kolbe 2011). The brand identity of Helsinki had traditionally been communicated as "the heart" of Finland, highlighted by beautiful nature and welfare state values (Hospers 2011). This brand identity had to be recreated in order to better appeal to the new generation of tourists (e.g. Poon 1994) demanding more differentiated tourism services of high quality.

As early as in 2002, HTCB started surveys Helsinki's image as a tourism destination, but the actual brand building process began in fall 2004 in co-operation with consulting companies and Helsinki Travel Trade. The initiative was taken by Kari Halonen, Marketing Manager of HTCB, who had met one of the main Finnish branding experts, Teemu Moilanen during his studies. From the outset, the main aim of the branding work was to differentiate Helsinki from its competitors, Stockholm, Copenhagen, Tallinn and St Petersburg, especially for short city breaks (Halonen 2010, Halonen 2011a; Halonen 2011b). The work on Helsinki's brand as a destination was made secretly to begin with and mainly by two persons: Kari Halonen and the Congress Director of the tourism bureau, Leena Sipilä. Since only two persons were involved at the beginning, and it was perceived that the process could not take place in public, this stage of the process was difficult. Halonen (2011a) describes the process like this: "The start-up stage took a year, a very painful year mentally and physically".

During the first year of brand building the focus was mainly on preparing a new slogan for Helsinki. The old slogan, "All Points Considered", was unclear both for people in the tourism bureau as well as for distributors. In fall 2005 a new slogan, "Helsinki – a Gateway between East and West", was presented by Halonen and Sipilä to their colleagues during a seminar. This slogan was developed by the initiative takers of the branding process and gained credibility thanks to a study conducted for an English travel magazine. According to this study, compared to other city break destinations in Europe Helsinki is a city culturally blended from both East and West (Halonen 2011a). The main idea behind this slogan is that Helsinki is a place in which Eastern exotica meets Scandinavian chic (HTCB 2011). The 700 years as a part of Sweden, followed by hundred years as a part of Russia prior to the independence in 1917, can strongly be seen in the city. The cultural heritage of the Russian era, was now seen in a more positive way, since this element was perceived to be a competitive advantage compared to Copenhagen and Stockholm (Kolbe 2011; Halonen 2011a). After several workshops the cathedral was chosen as the main symbol from a number of options, such as the famous Rock Church. As Halonen (2011a) explains: "The cathedral of Helsinki is like the Eiffel Tower for Paris or the Big Ben for London (Halonen 2011a)." Other attributes chosen to be used to describe the brand identity were archipelago, metropolis with human dimension, architecture, design, excellent service and a city full of events (HTCB 2011; Halonen 2011b).

In order to further enhance the interest of the private sector for the brand building work an annual competition, Best of Helsinki Awards, was launched in 2006. This campaign was one way to involve local stakeholders, such as distributors and local people in the branding process. Since 2006, companies which produce excellent tourism services supporting the Helsinki Brand have been awarded during a gala dinner at the end of each year (Halonen 2011a). One key attribute describing the brand identity of Helsinki as a destination was the service culture, and the Helsinki Award is one way to support this attribute. Further aims of this award ceremony are to make politicians more aware of the industry, and also to make the private tourism sector notice the need for product development (Halonen 2011a).

During 2006, the national tourism DMO, the Finnish Tourist Board, started to develop a brand for Finland as a destination. Kari Halonen was highly involved in this process, which ended up with a brand identity of Finland comprising 4 Cs: cool, contrasts, creative and credible. It is commonly perceived that the brand of Finland and that of Helsinki as tourism destinations support each other very well (Halonen 2011a, 2011b).

According to a survey conducted in 2007 by Copenhagen's DMO, Wonderful Copenhagen, comparing 14 cities in the Nordic and Baltic countries, Helsinki was seen as modern and trendy (Soininen 2008). This survey supported the branding work of HTCB, especially the idea of stressing Helsinki as a city characterised both by eastern culture and by Scandinavian chic. Another result of the survey was that people in Helsinki were perceived as being more approachable and exotic than people in other major Nordic cities (Halonen 2011a).

The brand identity of Helsinki was reorganised into three different groups of brand attributes, based on the survey. The first group of brand attributes comprises the core attributes of the Helsinki brand such as archipelago, events, design, Finnish cuisine, pocket sized city and the main symbol, the Cathedral. The second group of attributes is related to Helsinki's residents and efficient service culture, including a well-developed infrastructure. The third group is linked to the social brand, i.e. the flat hierarchy, equality, and the fact that the residents are easy to approach (HTCB 2011). Subsequently, there was a clear focus on linking a more human touch to the brand and thereby improving the brand personality.

Soon after the beginning the brand building process has been stable and work has been continued with limited financial resources, compared to the main competitors in the short holiday tourism market; Stockholm, Copenhagen and Gothenburg. A brand manual was developed and has been updated every year. A team, chaired by Kari Halonen is checking how the brand is portrayed in different electronic and printed media (Halonen 2011a).

The main marketing campaign, aimed at strengthening the image of Helsinki's brand, has been the Nordic Oddity Campaign. The main target groups of this campaign are couples without children, and families (Halonen 2011b). The core product of the campaign is the Nordic Oddity Brochure which was launched in 2004, but is continuously updated. It is offered in different languages, both in print and in digital form, downloadable from the Internet. The brochure includes information about different year-round attractions and activities in Helsinki, especially those of interest to younger people, such as open-air markets, cemeteries and cheap restaurants (Ventälä 2012).

4.4 Branding Helsinki for business

In 2006, a new organisation, called Greater Helsinki Promotion (GPH) was founded, financed by the city of Helsinki and its neighbouring cities (Espoo, Vantaa and the Regional Council of Uusimaa). The original mission of the organisation was to promote the Helsinki region as an international business location. This mission of GPH, however, is not so clear for the stakeholders. The leader of the organisation, Tatu Laurila (2011), describes the situation: "The mission of the GHP is not that clear to all our stakeholders. Especially those in tourism promotion were confused about our aims, thoughts and resources". Even though branding Helsinki as a concept has never been the main task of GHP, the idea of place branding has been there from the beginning. According to Laurila (2011), a strong brand is the most valuable tool for business promotion, but brand building is beyond the resources and competences of the GHP. For example, in the Annual Report (GPH 2011b, 4) it is stated that within the branding work the main task of GHP is to strengthen the international identity of Helsinki.

Laurila (2011) states that the GPH organisation would be interested in participating in some form of network aiming at building a strong brand of Helsinki, both as a business location and as a tourism destination, but GHP should not be the leader of the process. So far, though, no such network has been established. As Laurila (2011) stresses:"The fact that a clear brand building process does not exist is a pity, and even ridiculous" According to him, GHP has survived as an organisation by developing a brand definition, key messages and promotional tools of their own in order to be able to do their main work, i.e. promoting Helsinki for business. The main brand name is the Helsinki Business Hub which highlights the metropolitan area as a whole rather than the separate cities. On the other hand the organisation suffers from the fact that no strong Helsinki identity exists, nor is it under work. Even though some small steps have been taken, co-operation at grass root level, for instance social capital sharing, is still challenging between the stakeholders of GHP. As long as a the stakeholders of GHP are competing more than co-operating, a strong common brand building process of the Helsinki metropolitan area remains challenging.

4.5 The process of building the place brand of Helsinki

Weaknesses

One of the main weaknesses related to the whole brand building process of Helsinki is the lack of political will and an lack of understanding of what branding is about among decisions-makers. Laurila (2011) states that brand wise Helsinki has become much stronger and is seen as cool and attractive both by tourists and trend leaders, but the brand is much better than the process. Laurila is sorry for the fact that the local municipalities and their leading organisations have not been able to put their assets together and build a strong Helsinki brand with a long term perspective. The reason for this, according to Laurila, is the focus on the needs of their own city rather than the Helsinki region as a whole. Decision-makers do not understand the importance of building the brand.

Laurila (2011) stresses that in business you have to be concrete doers and thinkers, and investments in a brand are sometimes fuzzy and offer too little return on investment. According to Laurila top decision-makers have many other important issues to think about. Thus investments in brands and improved accessibility by attracting new flight connections are seen as costs rather than as image improvements. Accessibility is a key issue in branding and decision-makers should be prepared to invest the money airlines require in order to start new connections from Helsinki (Laurila, 2011).

It is not only the decision-makers who have had difficulties in understanding of what branding is about. The key actors in tourism have showed surprisingly little interest in the destination brand of Helsinki. The problem with the travel trade is that they lack time and resources to actively participate in the brand building process as Halonen (2011a) points out: "The focus of the tourism service providers is more on selling their own services than on seeing the importance and advantage of a strong destination brand." Another weakness of the process has been internal struggles inside the HTCB. To begin with the majority of the personnel of the Helsinki Tourism and Convention Bureau did not really understand the process of destination branding. For them branding was like going to a trade show. For example, Kari Halonen was interested in developing an e-learning platform for training of different stakeholders to become good hosts for tourists visiting Helsinki. The cost would have been around 30,000 euros, which is very affordable according to Halonen, since they have 8000 stakeholders in the database and most of them are unable to meet. But the branding process, and maybe also Halonen himself, lacked credibility among the board members of HTCB and this e-learning platform could never be launched. Thus the brand distribution has not been very advanced (Halonen 2011a).

It appears that one of the main challenges related to the branding of Helsinki has been the lack of a common vision for what the brand building process could comprise in the two main organisations (HTCB and GHP). For instance in HTCB there was no understanding for why GHP had to develop its own visual design without discussing it with the representatives of tourism branders. Halonen (2011a) declared that "we have this constant unclear situation about the image. There is a battle between HTCB and GHP." On the other hand the manager of GHP, Tatu Laurila (2011), stresses that the two organisations have a good relationship, although a somewhat confused setup exists. Halonen (2011a) describes this as "friendly fighting". A clear vision of what is the real destination brand of Helsinki is lacking. To take and example, Halonen (2011a) does not like the idea of Helsinki being branded by GHP as an innovative city since no studies have been conducted to support this statement.

One concrete example of what happens when a common vision of the brand does not exist is that GPH created their own graphic design covering Vantaa, Espoo and Helsinki, which was not even approved by the local councils. But many departments in Helsinki took this as the Helsinki brand abroad. For example, the Event Helsinki office focusing mainly on attracting sports events to Helsinki and part of the same economic department as the tourism bureau, are using the design of GHP in their promotion (Halonen, 2011). On the other hand HTCB is the most important actor marketing Helsinki internationally and they host 400 foreign journalists visiting Helsinki each year.

When it comes to the whole idea of place branding, the different offices in Helsinki work on their own without mutual communication, and tourism is separate from other trades (Kolbe 2011). On the other hand Laurila (2011) does not see any significant contradictions between the processes, but no synergy and no co-operation between the GHP developing the place brand of Helsinki and HTCB building the tourism brand of Helsinki exists. Very few shared visions can be seen behind the branding processes. Laurila highlights for instance the way how Visit Copenhagen and Copenhagen Capacity work together in the same building, having a total of 150 people doing practically the same work as only a couple of people are doing in GHP and in HTCB.

Branding Helsinki as a tourism destination has mostly been a process lead by active civil servants, which has been based on intuition rather than on strategy (Halonen 2011a; Kolbe 2011). However, the manager of the Helsinki destination branding process reflected in a recent article that there is a need for managing a destination brand, and management of the brand should be seen as a core competence of all organisations. On the other hand one of the branding experts interviewed stated that "nobody leads the process in Helsinki. I do not even know whether it is necessary" (Kolbe 2011).

4.6 Positive signals

One positive progress in the brand building process deepening the understanding for how important brand building is also to decision makers, was the possibility of Helsinki to host the Eurovision Song Contest after the victory of Lordi in 2006. This was the first time many different stakeholders worked together to create an urban carnival. Different stakeholders invested a lot of resources in order to create an event to be remembered all over Europe. During this time many decision makers got an understanding of how important events are for the brand (Laurila 2011).

For decision makers it is important to understand that with exposure and media attention, for instance during big sports or other large international events, the brand can be improved (Laurila 2011). World Design Capital in 2012 is another project that brought different actors together around joint aims and goals (Kolbe 2011; Laurila 2011). Politicians love media attention and thus it is easier to get them involved in common work around special occasions rather than on a long term brand building process (Laurila 2011).

5. Conclusions

When analysing Helsinki's branding process with the starting point from the model by Moilanen and Rannisto (2009) presented in Figure 2 the following conclusions can be drawn.

Start-up phase

The start-up phase of this tourism branding process was special. The driving forces behind the process were two employees at the Helsinki City Tourist and Convention Bureau, whilst other stakeholders were not included at all. This was not ideal from the networking point of view, and valuable chances of enhancing the visibility of the project were missed. Decision makers do not understand the branding process, according to Halonen (2011a), and all actors should work together to combine the limited resources for optimal effects in international markets. Maybe the starting point should have been to involve the mayor in the brand building process from the beginning, since then the process would have been given legitimacy from the beginning.

Research stage

Research carried out during the brand building process was mainly based on studies conducted by students at university tourism programmes, at Bachelor and Master levels. Most of it were analyses related to the image of the destination, both externally and internally. Consultants were also used to help identify the core of the brand. Studying the process now, it seems like even more research should have been carried out among internal stakeholders in order to get a more comprehensive view of what could be stressed in the brand identity.

Forming brand identity

Several workshops were organised to test and choose the elements of the brand identity. Following a number of workshops the cathedral was chosen as the main symbol for Helsinki's brand. Other attributes chosen to be used as part of the brand identity were archipelago, metropolis with human dimension, architecture, design and a city full of events. As part of the strategic plan for communicating the brand, an annual competition was launched to award companies that produced products that support the brand.

Execution

The execution of the brand has been implemented in international markets mainly through the Nordic Oddity campaign focusing on attracting young couples without children and families to experience a trendy Helsinki full of life and offering activities all the year round, 24 hours a day. A fact is though that compared to its main competitors, such as Stockholm, Gothenburg and Copenhagen, HTCB's budget for marketing communication has always been small (Halonen 2011b).

Follow-up

Once the brand has been implemented, follow-up is needed on the development of service processes and physical infrastructure. One way of following up the brand image has been the series of brand image research that students at the HAAGA-HELIA University of Applied Sciences have carried out since 2002. The first stage of the work was to conduct a content analysis on what media in chosen countries wrote about Helsinki as a travel destination. The students come from dozens of countries so a wide variety of markets could be covered. In 2010, a new element was added to this follow-up work when students also started to collect data in chosen markets by interviewing locals on the brand image of Helsinki. Over 10 markets were covered in 2011, and 2012.

Finally, the quote of Laurila (2011) sums up the branding process in Helsinki: "The brand itself is stronger than the process. Different interests, backyard fighting and local politicians focusing on their own region rather than the common good make the situation challenging"

References

Aaker, D. A. (1991) *Managing brand equity: capitalizing on the value of a brand name.* New York, Free Press.

Aaker, D. A. (1996) *Building strong brands.* Bath, The Bath Press.

Aaker, D. A. and Joachimsthaler, E. (2002) *Brand leadership*. London, Free Press.

Anholt, S. (2010) Why National Image Matters. In: UNWTO (ed.) *Handbook on Tourism Destinations Branding*. Madrid, World Tourism Organisation.

Blain, C., Levy, S. E. & Ritchie, B.R. (2005) Destination Branding: Insights and Practices from Destination Management organisations. *Journal of Travel Research*, 43 (May), 328–338.

Boo S., Busser, J. & Baloglu, S. (2009) A model of customer-based brand equity and its application to multiple destinations. *Tourism management*, 30 (2), 219–231.

Bornhorst, T., Ritchie, J. R. B. & Sheehan, L. (2010) Determinants of tourism success for DMOs and destinations: An empirical examination of stakeholders' perspectives. *Tourism Management*, 31 (5), 572–589.

Cai, L. A. & Hobson, S. P. U (2004) Making hotel brands work in a competitive environment in *Journal of Vacation Marketing*, 10 (3), 197–208.

Garcia, J. A., Gomez, M and Molina, A. (2012) A destination branding model: An empirical analysis based on Stakeholders. *Tourism Management*, 33 (3), 646–661.

Gilmore, F (2007) Brand Shanghai: harnessing the inner force of people and place. In: Morgan N., A. Pritchard, A & Pride, R (ed.): *Destination Branding- Creating the Unique Destination Proposition*. Kings Lynn, Elsvier, 169–184.

Greater Helsinki Promotion (2011a): *Revealing new business opportunities in Helsinki*, Brochure.

Greater Helsinki Promotion (2011b): *Annual Report 2010*. Helsinki, GPH.

Guion, S. 2002 *Triangulation: Establishing the Validity of Qualitative Studies*. Available from: http://rayman-bacchus.net/uploads/documents/Triangulation.pdf. [Accessed Ocotber 8th 2014].

Hall, M. C. (2008) Santa Claus, Place Branding and competition. *Fennia*, 186:1, 59–65.

Halonen, K. (2010) *Helsinki destination branding for tourists*. http://www.youtube.com/watch?v=8nGm8_3S4dQ, [Accessed October 11th 2011].

Halonen, K. (2011a): Interviewed by: Lindroth, K. (10th Ocotber 2011).

Halonen, K. (2011b): Tallinna?...Onko se jotakin syötyvää? Brändin kehittämisestä meillä ja muualla. *Ikkunapaikka*, 8/2011, 5.

Hampf, A. & Lindberg-Repo, K. (2011) *Branding: The Past, Present, and Future: A Study of the Evolution and Future of Branding.* Hanken Working Papers 556. Available from: https://helda.helsinki.fi/handle/10138/26578 [Accessed 15th October 2011].

Hospers, H. (2011): City branding and the tourism gaze. In: Dinnie, D. (ed.) *City Branding Theory and Cases.* Hampshire, Palgrave, 15–37.

Helsinki Tourism and Convention Bureau (HTCB 2012) *Helsinki City Tourism Brand.* Available from: http://www.visithelsinki.fi/In_English/Trade/Why_Helsinki/Brand.iw3 [Accessed 21st January 2012].

Helsinki Tourism and Convention Bureau (HTCB 2014) History of Helsinki in a Nutshell. Available from: http://www.visithelsinki.fi/en/come/welcome-helsinki/history-helsinki-in-nutshell [Accessed 8th October 2014].

Jenkins, O. H. (1999): Understanding and measuring tourism destination images. *International Journal of Tourism Research*, 1, 1–15.

Kolbe, L. Interviewed by: Lindroth K, K. (18th November 2011).

Kotler, P. & Gertner, D. (2002) Country as brand, product and beyond: A place marketing and brand management perspective. *Journal of Brand Management*, Apr, 249–261.

Laurila, T. Interviewed by: Lindroth, K. & Holmberg, E. (16th November 2011).

Molinanen, T. & Rainisto S. (2009): *How to Brand Nations, Cities and Destinations: A Planning Book for Place Branding.* Hampshire, Palgrave.

Morgan, N. & Pritchard, A (2007): Meeting the destination branding challenge. In: Morgan N., A. Pritchard A, & Pride, R. (ed.) *Destination Branding Creating the unique destination proposition.* Kings Lynn, Elsvier, 59–78.

Pike, S. (2010) Destination Branding Case Study: Tracking brand Equity for an Emerging Destination between 2003 and 2007. *Journal of Hospitality & Tourism Research*, 34 (1), 124–139.

Pike, S. (2009) Destination brand positions of a competitive set of near-home destinations in *Tourism Management*, 30 (6), 857–866.

Pike, S. (2004): *Destination Marketing Organisations.* Oxford, Elsevier.

Pitt, L. F., Opoku, R., Hultman, M., Abratt, P., & Spyropoulou, S. (2007) What I say about myself: Communication of brand personality by African countries. *Tourism Management*, 28 (3), 835–844.

Poon, A. (1994) *Tourism, Technology and Competitive Strategies.* Wallingford, Cabi.

Ritchie J. R. & Ritchie, R. J. B. (1998) *The branding of tourism destinations, Past Achievements and Future Challenges*, Report prepared for the 1998 Annual Congress of the International Association of Scientific Experts in Tourism in Morocco, September 1998.

Soininen, P. (2008) Matkailijoiden mielestä Helsinki on moderni ja trendikäs. *Matkailusilmä*, 4, 14–15.

Stake, R. (1995) *The Art of Case Study Research*. Thousand Oaks, Sage.

Tasci, A. D. A. (2011) Destination branding and positioning. In: Wang, Y. & Pizam, A. (Ed.), *Destination Marketing and Management: Theories and Applications*. London: CABI, 113–129.

Therklsen, T. & Gram, M. (2010) Branding Europe – Between Nations, Regions and Continents. *Scandinavian Journal of Hospitality and Tourism Research*, 10 (2), 107–128.

Usakli, A., & Baloglu, S. (2010) Brand personality of tourist destinations: An application of self-congruity theory. *Tourism Management*, 32 (1), 114–127.

Ventelä, S. (2012) Nordic Oddity –kampanja esittelee nuorekasta Helsinkiä. http://www.rantapallo.fi/matkailu/nordic-oddity-%E2%80%93kampanja-esittelee-nuorekasta-helsinkia/ [Accessed 17th May 2012].

Vuoristo, K.-V. (2002) Regional and structural patterns of tourism in Finland. *Fennia*, 180 (1–2), 251–259.

Xiao, H. & Smith, S. L. J. (2006) Cases studies in tourism research: A state-of-the-art analysis. *Tourism Management*, 27 (5), 738–749.

Yin, R. K. (2003) *Case study research: design and method*. Thousands Oaks, Sage.

Jarmo Ritalahti and Kaija Lindroth

Economic Impacts of Tourism in Eastern Uusimaa, Finland

1. Introduction

The province of Eastern Uusimaa is a popular tourism destination about 50 km east of Helsinki, the capital of Finland. Its main attraction is the medieval town of Porvoo with its old town centre and a grey stone cathedral from the 15th century. Other attractions include the small and idyllic town of Loviisa, the surrounding archipelago and living countryside.

This study on economic impacts of tourism in the province of Eastern Uusimaa was conducted in 2009 and used the Nordic model to measure the tourism income in the region. The Nordic model measures both tourism expenditure and revenues. In the case of Eastern Uusimaa, the tourism expenditure was estimated from the results of structured visitor interviews and surveys for local residents and second home owners. The tourism revenue study included an Internet-based survey of tourism and tourism-related companies, followed by statistical analysis.

According to the study the province of Eastern Uusimaa received over 1, 7 million tourists in 2009. The biggest visitor group was day-trippers followed by visitors staying with friends and relatives (VFR). The direct tourism revenue according to the Nordic model without value added tax (VAT) was 163 million euros and the total (direct + indirect) revenue (excluding VAT) amounted to 220 million euros. The turnover of tourism in the region makes the industry the third largest following the chemical and food industries. The number of jobs created by tourism in 2009 was over 1400.

2. Destination under Study

2.1 Eastern Uusimaa Region

The region of Eastern Uusimaa comprises ten municipalities with a combined population of nearly 100 000. Due to its long history, the region has a plethora of cultural and historical venues with old and valuable architecture. It also boasts associations with many of the best-known artists in the country – both past and present.

The centre of the province, Porvoo, has long traditions as one of the most popular domestic day-trip destinations due to its central role in Finnish history, its mediaeval town plan and wooden architecture from the 18th century. The rest of the region does not have significant tourism traditions but can be described as a stop-over for tourists travelling from the capital to the Lake District or as a day-trip destination for tourists using Helsinki as their base.

Thanks to its location close to the capital area, the region attracted 1, 7 million visitors in 2009. Over 80% of them were same-day visitors and Finns. Of the Finnish visitors 40% came from the province of Uusimaa, i.e. the capital area.

Most of the visitors spent a couple of hours in the area, attracted to the region by the cultural offering. Over 50% of the visitors travelled with their spouse and 77% had visited the area previously. (Ritalahti & Holmberg, 2011)

2.2 Eastern Uusimaa Tourism Strategy

The tourism strategy of the province for 2007 – 2013 defines the vision of the area as a destination that is accessible and well networked. According to the strategy, the aim is to make sure that the region will develop and offer authentic tourism products and experiences of high quality throughout the year both for individual tourists, groups of tourists, and business travellers. The key attractions of the province are the old towns of Porvoo and Loviisa, as well as the manor houses and the archipelago (Itä-Uudenmaan Matkailustrategia 2007–2013, 2006: 7.)

In 2006, when work on the 2007–2013 was on-going, the quantitative goals for tourism development in the region were set as shown in Table 1:

Table 1: Quantitative Goals for Tourism Development in Eastern Uusimaa Region for 2007–2013

Quantitative aims	Measure
Hotel occupancy rate	57%
Average stay	2 nights
Annual growth in Business travel	7 %
Services in archipelago	growth
Business activities in tourism	growth
Employment in tourism	growth
Tourist season	lengthening
Funding for tourism development	€ 2 M

Source: Itä-Uudenmaan matkailustrategia 2007–2013 (2006)

The qualitative goals were, as shown in Table 2:

Table 2: Qualitative Goals for Tourism Development in Eastern Uusimaa Region for 2007–2013

Qualitative aims	Measure
Improved quality of tourism services	use of quality systems
Improved customer satisfaction	tourist experiences
Event development	innovative events

Source: Itä-Uudenmaan matkailustrategia 2007–2013 (2006)

3. Methodology

3.1 Economic Impacts of Tourism

Tourism brings money to a region when people who come from outside buy services and goods in the destination. This tourism expenditure impacts on three levels: direct, indirect and induced levels. Companies that receive directly money from visitors respend it on goods and services, labour, taxes, licences and fees. One part of the money leaks out of the local economy as savings and imports. Indirect impacts refer to money flows from the tourism sector to other local businesses when purchasing services and goods. Induced impacts are salaries for local people, profits, rents and interests that are partly used in the local economy (i.e. Ratz & Puczko, 2002; Cooper, Fletcher, Fyall, Gilbert & Fanhill, 2008).

3.2 The Nordic Model

The development of the Nordic Model started at the end of the 1970's in a project funded by the Nordic Council of Ministers. The idea was to create a tool to measure the economic impacts of tourism at a regional level. The Nordic Model is based on the idea of using several methods in data collection.

The model consists of two processes that are carried out separetly. The demand side or expenditure method focuses on the money tourists or visitors spend in the target destination. The supply side or receipts method measures the money flowing into the enterprises at the destination. Direct tourism consumption and tourism income are compared to elicit an estimate of the economic impact of tourism in the destination under study (Paajanen 1994).

The variety of data collection methods depends on the aims of the project as well as the budget available. Methods to measure tourism expenditure include, for example tourist interviews, questionnaires for tourists, questionnaires for locals who accommodate friends and relatives, and questionnaires for non-resident second-home owners. Methods to estimate tourism incomes include surveys and interviews with local or regional enterprises that receive direct or indirect tourism income. Furthermore, statistical analyses of secondary data are also used to support empirical data directly from the industry.

According to the traditional Nordic Model process, the research to measure the economic impact of tourism in Eastern Uusiamaa can be divided into two parts. The methods to measure tourism expenditure were tourist interviews, and surveys with local residents and non-resident second-home owners. The method to collect data about tourism receipts was originally a survey sent to the enterprises operating directly or indirectly in tourism sector. However, due to the low number of responses from the enterprises, the final analysis is based on regional statistics.

4. The Research Process

4.1 Research Problem and Aims

The aim of the survey was set as: "defining the income or amount of money tourists spent in the province in 2009, as well as the number of jobs created by tourism".

Other interesting issues in the survey were the demographic background of the visitors, such as their home region or country, main purpose of the visit, type of trip, sources of information, services and goods purchased, attractions visited, nights spent, and booking or purchasing channel. All of this draws a clearer picture of tourists to the province.

Surveys on the economic impacts of tourism give concrete information about tourists' activity patterns and what is actually happening in the tourism industry in the province. The main aim of this survey is to help regional tourism stakeholders, especially decision makers to deepen their understanding of the importance of tourism as an industry and to act accordingly.

The limitations of this survey are linked to the multiplier effect of tourism income, lack of the study of leakages and induced income. The multiplier effect could not be calculated because the respondents, tourism enterprises, were not able to estimate how much they purchase products and services from other entrepreneurs. Therefore, an average multiplier was used instead of one based on the results oft he survey. Leakages and induced income were not regarded significant by the commissioner of this survey, and were subsequently left out.

4.2 Tourism expenditure

The tourism expenditure was measured by three different tools. The personal structured interviews of visitors were conducted between January and October 2009. They were developed in three languages, (Finnish, Swedish and English) in the towns of Porvoo and Loviisa, two most important destinations of the province. The size of the sample was 2742, and most of the interviews were conducted during the summer months of June, July and August, representing the peak of the tourism season in the province. The interviews were conducted daily in Porvoo and Loviisa in June and July, and during weekends in August. During other seasons tourists were interviewed only in the old town of Porvoo, which is the most visited attraction in the province. The interviews took place on the streets or other locations close to the most visited sites in the destinations. The interviewees were selected by random sampling, interviewers stopped tourists passing by and interviewed the those who fit the selection criteria (being a "tourist" or a "visitor").

When mining the data and analysing the results tourists were divided into the following six groups according to their type of accommodation in the province:

1. Tourists staying in hotels and motels
2. Tourists staying on camping sites, in hostels, B&B's or boarding houses
3. Tourists staying in boats in guest harbours
4. Tourists staying in their own second homes
5. Tourists staying with friends or relatives
6. Day-trippers

Data was also collected through postal surveys for second-home owners and local residents in the province. A total of 2000 questionnaires in Finnish or Swedish based on their mother tongue were sent to the second home owners in different municipalities by random sampling. The posted numeber of questionnaires followed the number of second-homes in the respective municipalities. The questionnaires with covering letters and reply-paid envelopes were sent in one mailing in October 2009. The response rate was 30%, i.e. 631 questionnaires were returned. This can be described as an excellent result.

In November 2009, 2970 questionnaires in Finnish or Swedish were sent to the local residents with the same procedure than the second-home owners. The aim of 500 usable responses was almost reached with the response rate of 16% (468). Respondents to both postal surveys were encouraged to participate by giving them an opportunity to take part in a prize draw with gifts from regional tourism service providers.

4.3 Tourism income

The original goal was to collect the data on the tourism income through a tourism enterprise survey on the Internet. The enterprises were chosen according to the classification of the Tourism Satellite Account (TSA) designed to measure the goods and services purchased by visitors. The characteristic goods and services in tourism are defined as:

1. Accommodation services
2. Food and beverage serving services
3. Passenger transport services
4. Travel agency, tour operator and tourist guide services
5. Cultural services
6. Recreation and other entertainment services
7. Miscellaneous services

The miscellaneous products are connected products like services such as fuel and non-specific products, such as retail in non-specialized stores.

The first Internet survey was sent to 84 enterprises in December 2009. Due to thelow number of responses, a second round was done at the beginning of January 2010, and a third at the end of January 2010. In spite of these three rounds the number of responses remained so low that the results could not be used as a main source of data for this research.

Thus the tourism income study was conducted as a statistical study using data based on the statistics of the different sectors of the economy in 2008. The figures of 2008 were converted to estimates of 2009 by using different coefficients of change between these two years.

5. Results

In 2009, two years into the new strategic period, a total of 160.812 overnight stays were registered in accommodation establishments and guest harbours in the area. Furthermore, nearly two million overnights were spent with friends and relatives while just over two million nights were spent in private second homes. The goal set in 2006, a 2-day average stay, had not been reached yet. Direct tourism revenue in 2009 was 163 million euros and total tourism revenue (excluding VAT) amounted to 220 million euros. Tourism employed directly 1 400 persons. The total tourist expenditure in 2009 was divided between the various segments as shown in Table 3:

Table 3: Tourist Expenditure in Eastern Uusimaa in 2009

Visitors staying with friends and relatives	€ 73 893 000
Same-day visitors	€ 58 520 000
Visitors in private second homes	€ 44 785 868
Hotel guests	€ 14 710 664
Visitors in guest harbours	€ 1 638 826
Other accommodation	€ 1 537 006
Total	€ 190 036 446

Source: Ritalahti & Holmberg (2011)

6. Discussion

The study on the economic impacts of tourism in the_Region of Eastern Uusimaa in 2009 coincided with the recession in world economy. This downturn may have had an impact on consumer behaviour in this region as well. However, 2009 marked the celebration of the Diet of Porvoo in 1809. The happenings that were staged in the region to commemorate this historic event may have compensated for the possible loss of revenue from abroad by bringing in an increased number of domestic visitors.

According to this study, the economic impact of tourism in the region is considerable. Direct economic gain in 2009 amounted to € 163 million (excluding VAT), whereas direct and indirect revenue together (excluding VAT) was € 220 million. This turnover has considerable significance in the region. The significance is indicated by the fact that nearly 1 400 people were employed in tourism in 2009.

Visitation to Eastern Uusimaa takes many forms but the biggest group consists of day-trippers. This dominating role of day-trippers, 1 100 000 out of the total of 1 700 000 visitors, is a distinctive factor to be considered when comparing the results with other regions. This category of tourists brought in € 53 million, whilst the segment bringing in most revenue, tourists staying with friends and relatives, brought in € 73 million. An average day-tripper spent € 48–61 per day, only 10 euros more than those staying outside registered accommodation, whereas hotel guests spent on average € 132, 34. This highlights the importance of working towards reaching the strategic goal of longer stays in the destination.

The findings of the expenditure side of the survey can be considered reliable whilst the income survey raises a lot of questions. On the whole, the implementation of the ambitious tourism strategy of the region relies heavily on dynamic entrepreneurs. However, judging by the poor rate of response to the income survey, the entrepreneurs do not seem to realise the value of research into the economic impacts of tourism. If they cannot be bothered to respond to the survey and, furthermore, if they are not able to estimate the share of tourism in their revenue, how can they be expected to develop innovative tourism products of high quality to meet the needs of future visitors? Is it the case that tourism businesses in the region are small lifestyle entrepreneurs who are not interested in growth and hence do not see the value of research? Or do they think that local decision makers are not keen on developing the industry, hence why participate in research? Could perhaps different research methods have been used to reach a better result?

The survey was carried out using the Nordic Model which offers various ways to approach the subject. However, this flexibility has its advantages and disadvantages; one benefit is that it can be applied to a variety of destinations. The possible approaches again lead to problems of comparing the survey results from different areas. (Jennings 2001)

There are a number of other features as well in the Nordic Model that are open to criticism: large empirical data, lack of statistical data on day-trippers, numbers of tourists staying in non-commercial establishments, interviewees' ability to estimate their expenses, the large number of stakeholders in destinations, categorising, sampling, and data collection techniques being some of the possible flaws. Large empirical data may be troublesome to handle and analyse. Day-trippers again as a target group are challenging because they do not necessarily visit places where visitor statistics are in place. Therefore there may not be much data on how they spend their money. The same is true of visitors staying overnight in private homes.

A further challenge comes when interviewing visitors about their use of money. People do forget how much they spend on various items, what they bought before the trip using a credit card, or they may simply give answers they think are expected of them. The stakeholders involved in surveys are another consideration as they have their own opinions and interests that may conflict with the interests of the destination as such. Furthermore, categorising and sampling within the field of tourism never cease to cause discussion as the definitions of tourism per se and what is included in it are a matter of debate. This leads to serious issues about how to best collect data and how reliable it is. (Kauppila 2001)

The combination of the above-mentioned weaknesses and the lack of precise definitions results in researchers in different regions using varying definitions, which in turn makes it difficult to compare results. One solution might be to aim at longitudinal studies where the same definitions and questionnaire techniques are used systematically over a longer period of time to allow for comparison. Due to recent changes in regional administration, i.e. Eastern Uusimaa becoming part of Uusimaa on 1 January 2011, the following study on economic impacts will not only cover the destination of Eastern Uusimaa but the entire Uusimaa province. Part of the funding for a similar survey has been guaranteed and it is expected to be carried out in 2013. It remains to be seen how it compares to current research and what changes in research methods will take place.

References

Cooper, C., Fletcher, J., Fyall, A., Gilbert, D. & Wanhill, S. (2008). *Tourism principles and practice* (4th ed.). Harlow, Pearson Education Limited.

Itä-Uudenmaan Matkailustrategia 2007–2013 (2006). Porvoo, Eastern Uusimaa Regional Council.

Jennings, G. (2001). *Tourism research.* New York, John Wiley & Sons.

Kauppila, P. (2001). *Matkailun aluetaloudelliset vaikutukset: pohjoismaisen mallin matkailijatutkimukset.* Naturpolis Kuusamo, tutkimuksia 3/2001.

Paajanen, M. (1994). *The economic interaction between tourists and enterprises.* Helsinki, The Helsinki School of Economics and Business Administration, B–142.

Ratz, T. & Puczko, L. (2002). *The impacts of tourism*: *An introduction.* Hämeenlinna, Häme Polytechnic.

Ritalahti, J. & Holmberg, E. (2011). *Itä-Uudenmaan matkailun tulo- ja työllisyystutkimus 2009.* HAAGA-HELIA tutkimuksia 3/2010. Helsinki, HAAGA-HELIA University of Applied Sciences.

Table of Figures

The challenges of destination branding – a case study of Helsinki

Author Bios

Kristian Bjørnstad is a Ph.D-candidate at the Norwegian University of Life Sciences and holds the position as secretariat director of the Norwegian Parks Association. He has years of experience as development advisor for regional parks and world heritage areas in Norway, and has been central in developing a national system of regional parks in Norway. He is currently actively building a new Nordic network of regional parks. Research interests focus on park development processes, European regional parks, protected areas and rural development.

David Di Tota completed his Bachelor of Commerce in Hospitality and Tourism Management at Ryerson University in Toronto (Canada) and Auckland University of Technology in Auckland (New Zealand) after graduating with Honours from the Small and Medium Enterprise Management program at Algonquin College in Ottawa (Canada). He spent over eight years working at the operations level for a number of luxury hotel brands in Ontario (Canada), and in early 2013 he began his professional career as a Management Consultant with Carpedia International, focusing predominantly on efficiency management within the hotel and restaurant industry.

Dr Bernd Eisenstein is Professor for International Tourism Management at Westcoast University of Applied Sciences in Heide, Germany, and since 2006 he has been the director of the Institute for Management and Tourism (IMT) of the Westcoast University of Applied Sciences. He completed his Dipl.-Kaufmann and Dipl.-Geograph at Trier University where he subsequently completed his doctorate, supervised by Prof. Christoph Becker. Bernd had lecturing positions at Trier University and the University of Applied Sciences Worms (Germany), and has conducted and coordinated numerous consultancy and research projects, in particular in the fields of destination management, tourism demand and strategic tourism management.

Eva Holmberg (Lic. Econ.) is a senior lecturer of tourism and methods at Haaga-Helia University of Applied Sciences in Porvoo, Finland. She teaches at both bachelor and master levels in courses such as brand management, responsible tourism, and destination management. Her research interests are mainly related to destination management, branding and responsible tourism development.

Dr John Hull is an Associate Professor of Tourism Management at Thompson Rivers University in British Columbia, Canada and a visiting professor at Buskerud and Vestfold University College, Norway and at the Harz University of Applied Sciences, Germany. He is also a member of the New Zealand Tourism Research Institute. His research addresses the sustainability of tourism planning and destination development in peripheral and rural regions. Specific areas of focus include cruise tourism, creative tourism, geotourism, culinary tourism, mountain tourism, polar tourism, and wellness tourism.

In addition to his academic experience, John has consulted for numerous public and private agencies worldwide. Internationally, past clients include the UNWTO, UNCBD, UNEP, UNESCO, Nordic Council, European Tourism Research Institute, European Central Development Programme, Office of North Atlantic Agreement, Icelandic Tourism Research Centre, Greenland Tourism and Business Council, Faroe Island Tourist Board, New Zealand Service IQ and the South Pacific Tourism Organisation. In Canada John has consulted with the Canadian Tourism Commission, Parks Canada, Tourism Atlantic and for the provinces of Newfoundland and Labrador, Quebec, Alberta, and British Columbia. He is co-editor of a book on Mountain Tourism (CABI).

Alexander Koch completed his Bachelor of Business Administration (BBA hons) in Leisure and Tourism Management at the University of Applied Sciences Stralsund (Germany) in 2009, with distinction. Since then he has been employed as a member of the project team of the Institute for Management and Tourism (IMT) of the Westcoast University of Applied Sciences in the fields of market research and consultancy.

Anne Köchling (MTM, Dipl.-Kffr.) studied International Business Studies at the University of Paderborn (Germany) and the University of Santiago de Compostela (Spain). After finishing her degree in Business Administration she completed a Master of Tourism Management at the Freie Universität in Berlin. In 2005, Anne started her professional career in destination management at the German National Tourist Board in Frankfurt, and from 2007 until 2009 she worked at the tourism organization for the federal state of Schleswig-Holstein, where she was in charge of marketing for the destination. Since 2010 she has been with the Institute for Management and Tourism (IMT) of the West Coast University of Applied Sciences in Heide/Holstein (Germany), where she is currently in charge of several market research and consulting projects in the field of destination management.

Dr Martinette Kruger obtained her PhD in 2010, where she critically evaluated market segmentation at arts festivals in South Africa. During her academic career at the North-West University in Potchefstroom (South Africa), she has obtained numerous awards including the ABSA Bronze Medal winner for the best master's student at the North-West University in 2010, and in 2013 she was honoured as the most productive junior researcher in the faculty of Business and Management Sciences. She is involved in numerous research projects conducted by TREES (Tourism Research in Economic Environs and Society) at selected national parks, arts festivals and events in South Africa. She has published numerous articles in national and international journals, and is part of the editorial board for various national and international journals. She is currently working as an Associate Professor as part of the tourism research focus area TREES.

Pascal Languillon is the director of www.travelfortheplanet.com, a consumer website dedicated to the promotion of sustainable travel options around the world. He is also a travel writer and the author of many articles and guidebooks about high-end ecotourism, including the coffee-table book Ecochic, published in 2010, and a Lonely Planet guidebook published in French. Pascal founded the French Ecotourism Association in 2007, and worked as a tourism consultant to help establish sustainable tourism initiatives around the world. He relocated to New Zealand in 2010, where he now works as a tourism marketer.

Kaija Lindroth (M.A.) is the Degree Programme Director of the two international Bachelor programmes at Haaga-Helia University of Applied Sciences, Porvoo Unit (Finland): International Sales and Marketing, and Tourism. She is responsible for the internationalisation of the Haaga-Helia Porvoo Campus degree programmes, and passionate about developing professional networks for the Porvoo Campus needs both towards the industry and academia.

Her research interests include network studies among small tourism entrepreneurs, social impacts of events, and branding. She has been actively involved in the community of tourism researchers both in Finland (Finnish Society for Tourism Researchers) and internationally (TTRA European Chapter).

Dr Michael Lück is an Associate Professor of Tourism in the School of Hospitality and Tourism, and Associate Director for the coastal and marine tourism research programme at the New Zealand Tourism Research Institute, both at AUT University, New Zealand. He is founding co-chair of the International Coastal & Marine Tourism Society (ICMTS). Michael has more than 10 years work experience in the tourism industry and his research interests include (marine) wildlife tourism, the cruise industry, ecotourism, the impacts of tourism, and aviation. He has published in a number of international journals, is founding editor-in-chief of the academic journal Tourism in Marine Environments, and Associate Editor of the Journal of Ecotourism. Michael has edited or co-edited ten volumes on ecotourism, marine and polar tourism, events and low cost airlines, as well as the Encyclopedia of Tourism and Recreation in Marine Environments (CABI), and co-authored the introductory text Tourism (CABI).

Dr Tomas Pernecky is a senior lecturer in the School of Hospitality and Tourism at the Auckland University of Technology, Auckland, New Zealand. His research focuses mainly on knowledge production of and theoretical inquiries into events and tourism, epistemology in qualitative research, hermeneutic phenomenology, sustainability, and the critical exploration of issues surrounding peace. Tomas is also the co-founder and advocate of post disciplinary approaches to the study of tourism (www.postdisciplinary.net).

Dr Brooke Porter joined Auckland University of Technology/New Zealand Tourism Research Institute in 2011 as a PhD student, and graduated in December 2014. Brooke completed a Bachelors of Science in Marine Biology at Florida Tech in Melbourne, Florida and relocated in 2001 to work with a leading marine ecotourism NGO in Maui, Hawai'i. Becoming involved in the post-tertiary training of naturalists, Brooke returned to university to gain formal training in applied educational theory. She holds a Masters in Education from Chaminade University, Honolulu, Hawaii.

Her PhD explored the integration of rural marine tourism into the fishing communities in less developed nations. The impetus for this research began while on assignment in Eritrea, Africa coordinating an internationally funded Fisheries Development Project. Her research seeks to provide developing countries with alternative development strategies for marine resources. Brooke's other areas of interests include fisheries management, ecotourism, interpretation in wildlife tourism, voluntourism, service learning and community outreach.

Jarmo Ritalahti is a principal lecturer of tourism at Haaga-Helia University of Applied Sciences. His main research interests in tourism focus on intermediation and business tourism, and pedagogical development and pedagogy in higher education. He was responsible of the new pedagogical approach and curricula development at Haaga-Helia's Porvoo Campus project in 2007–2011. Jarmo is also working on regional tourism research and development projects.

Jonathon Spring is a lecturer in the Tourism Management Programme at Auckland Institute of Studies, and a PhD candidate at AUT University, both in Auckland, New Zealand. The topic of his PhD study is on visitor-guide interactions during guided wildlife tours conducted by the Supporters of Tiritiri Matangi in New Zealand and Pacific Whale Foundation in Hawai'i. His research interests are in guide training, informal learning, and ecological restoration. Jonathon is also a volunteer guide on Tiritiri Matangi.

Xiaoshu (Sue) Sun was born in Chongqing, China. She joined Auckland University of Technology in 2012 as a student of the Master of International Tourism Management programme. In 2011, she completed a Bachelor Degree of Tourism Management at Chengdu University of Technology in Chengdu, China. Her masters dissertation explored the role of New Zealand whale watching operators in ecotourism through analysing the contents of their websites. The research identified contributions the New Zealand whale watching operators make for the conservation of the marine environment and development of the local communities. Sue's other areas of interests include wildlife tourism, volunteer tourism, responsible tourism and dark tourism.

Dr Peet van der Merwe was born in Kuruman, South Africa and is currently living in Potchefstroom South Africa. He is a professor and researcher in tourism management at North-West University: Potchefstroom Campus, School for Business Management. As researcher he also forms part of the NRF research unit TREES (Tourism Research in Economic, Environs and Society). Peet obtained his degree in 1994, honours degree in 1995 and master's degree in 1999. He completed his PhD (2004) in tourism management. His field of expertise lies in wildlife tourism, hunting tourism and ecotourism (sustainable tourism). Peet has widely published, such as peer reviewed journal articles, popular articles, project and research reports, books and book chapters. He has presented at numerous domestic and international conferences, and supervised 15 masters and PhD students.

Dr Jan Velvin is an Associate Professor in the Department of Industrial Economics and Technology Management, at Buskerud and Vestfold University College, Norway. He also holds a position as Director at the Center of Tourism Management at Buskerud and Vestfold University College and Visiting Associate Professor at the Scandinavian College in Brazil. Jan has worked on issues in tourism since 1994. His research interests are in the areas of protected areas and rural development, second home tourism, tourism statistics and service management.

Schriftenreihe des Instituts für Management und Tourismus (IMT)

Herausgegebenvon der Fachhochschule Westküste

Die Bände 1-6 sind im Martin Meidenbauer Verlag erschienen und können über den Verlag Peter Lang, Internationaler Verlag der Wissenschaften, bezogen werden: www.peterlang.de.

Ab Band 7 erscheint diese Reihe im Verlag Peter Lang, Internationaler Verlag der Wissenschaften, Frankfurt am Main.

Band 7 Anja Wollesen: Die Balanced Scorecard als Instrument der strategischen Steuerung und Qualitätsentwicklung von Museen. Ein Methodentest, unter besonderer Berücksichtigung der Anforderungen an zeitgemäße Freizeit- und Tourismuseinrichtungen. 2012.

Band 8 Wolfgang Georg Arlt (Ed.): COTRI Yearbook 2012. 2012.

Band 9 Michael Lück / Jan Velvin / Bernd Eisenstein (eds.): The Social Side of Tourism: The Interface between Tourism, Society, and the Environment. Answers to Global Questions from the International Competence Network of Tourism Research and Education (ICNT). 2015.

www.peterlang.de

www.ingramcontent.com/pod-product-compliance
Ingram Content Group UK Ltd.
Pitfield, Milton Keynes, MK11 3LW, UK
UKHW021014120926
8556IPUK00040B/1129